AUDIENCES for PUBLIC TELEVISION

HE
8700
.7
.A8
F69

AUDIENCES for PUBLIC TELEVISION

RONALD E. FRANK
and
MARSHALL G. GREENBERG

Poynter Institute for Media Studies
Library

JUN 26 '87

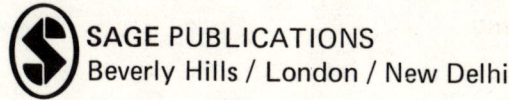

SAGE PUBLICATIONS
Beverly Hills / London / New Delhi

Copyright © 1982 by Sage Publications, Inc.

All rights reserved. No part of this book may be reproduced or utilized in any form or by any means, electronic or mechanical, including photocopying, recording, or by any information storage and retrieval system, without permission in writing from the publisher.

For information address:

                SAGE Publications, Inc.
                275 South Beverly Drive
                Beverly Hills, California 90212

SAGE Publications India Pvt. Ltd.	SAGE Publications Ltd
C-236 Defence Colony	28 Banner Street
New Delhi 110 024, India	London EC1Y 8QE, England

Printed in the United States of America

Library of Congress Cataloging in Publication Data

Frank, Ronald Edward, 1933-
   Audiences for public television.

   Bibliography: p.
    1. Television audiences—United States. 2. Public television—United States. I. Greenberg, Marshall G., 1935-   II. Title.
HE8700.7.A8F69 1982   384.55'44   82-16754
ISBN 0-8039-0764-8

FIRST PRINTING

to LLOYD N. MORRISETT
Thanks for the opportunity to fill a prescription!

Contents

Preface	11
Acknowledgments	15
Chapter 1: Perspectives	17
Chapter 2: Study Design	31
Chapter 3: The Audience Interest Segmentation	45
Chapter 4: Public Television Viewing Behavior	79
Chapter 5: Media Usage Among Above-Average PTV User Segments	101
Chapter 6: Media Usage Among Below-Average PTV User Segments	121
Chapter 7: Public Television Program Preferences	155
Chapter 8: Public Television Funding	173
Chapter 9: Minority Audiences	187
Chapter 10: Summing Up	209
Appendix	225
References	229
About the Authors	231

AUDIENCES for PUBLIC TELEVISION

Preface

In 1974, The John and Mary R. Markle Foundation sponsored a one-day seminar to discuss research strategies for identifying "special interest" audiences for public television (PTV). The best statement of the need for PTV special interest audience identification and development at the time was, and still is, provided by Dr. Lloyd Morrisett, president of the foundation, in a paper originally contained in its 1972-1973 annual report entitled, "Rx for Public Television."

That meeting stimulated us over the next two years to undertake an exploratory, hypothesis-generating study and a rather extensive pilot project to test a large-scale national survey research methodology. Our experience in these two efforts convinced us that to really understand audiences for PTV it was crucial to develop a framework of television viewing in general. Toward this end, we recognized the importance of examining television usage in the context of a viewer's interests in a variety of leisure activities and subjects. By doing so we felt we could contribute useful information for audience development to those concerned with commercial television as well as to those concerned with PTV.

This recognition led to the design and implementation of a large-scale national survey, the PTV findings of which are reported in this book. It is an investigation of the functions of commercial television, as well as of public television and other media, from the viewpoint of their respective audiences. Our primary purpose is to develop new insights into the ways in which audience members use television, both commercial and public, and other media in their daily lives.

Because we thought that the general context in which public television was used was crucial to understanding it, we initially focused our analysis on overall television viewing behavior. This resulted in a prior book entitled *The Public's Use of Television* (Sage, 1980), which contained only one chapter exclusively devoted to a discussion of PTV. Upon completion of that book, both the Corporation for Public Broadcasting and The John and Mary R. Markle Foundation expressed interest in extending our efforts to analyze the data further and relate our findings specifically to PTV. The present book is designed to achieve that goal.

Since this and our previous book derive their data bases from the same survey and share a common segmentation analysis, there is necessarily some overlap between them in the discussions of methodology and even in some of the results and conclusions. While we have tried to make the present book self-contained, we have abbreviated our earlier discussion of methodological details and have eliminated much of the material included in the appendices to our previous book. The reader interested in a copy of the entire survey questionnaire, or in detailed data on viewing of individual programs on commercial television by interest segments, will find them in those appendices, along with detailed usage data on other media.

Our intended audience comprises academic, industry, and government professionals interested in better understanding the role of PTV. Included are:

(1) Those directly associated with the process of generating, evaluating, and choosing television programming material. This includes people responsible for program development and research, especially in public television networks and stations, along with people in similar positions in television production companies.
(2) Those in marketing organizations (manufacturers), public relations firms and advertising agencies that might be interested in funding the development and/or airing of PTV programming.
(3) Faculties and student bodies in schools of communication, journalism, and business (especially those interested in advertising and mass media).
(4) Those in special segments of the general public. There are a number of organizations, such as Action for Children's Television (ACT), whose members might think our findings of interest.

Given these diverse audience objectives, we have reported the findings in what we hope to be a readable style, irrespective of the reader's background or skills (given the inherently technical nature of the research methodology).

The reader who is interested in a rather abbreviated overview of the project and our conclusions can obtain it by reading Chapters 1 and 2, the segment sketches in Chapter 3, and finally, Chapter 10.

Many studies have been conducted to examine television viewing behavior, though very few have concerned themselves with an in-depth investigation of PTV. While our study describes the behavior of audiences for PTV and other media, that is not its principal goal. Our intent is to try to understand this behavior better by examining its relationship to other audience characteristics, especially individual patterns of leisure interests and activities and the psychological needs they satisfy. We have also attempted to use this understanding to develop suggestions as to changes in PTV programming that might serve to increase its audience reach without sacrificing its standards of excellence.

—R.E.F.
—M.G.G.

Acknowledgments

The research on which this and our previous book are based derives from a series of meetings held in 1974 that stimulated our thinking about the possibility of applying market segmentation techniques to the study of public television audiences. Since that time the people who have influenced our ideas about the project or assisted in its implementation number in the hundreds. To all of those who offered their support and critical comments in meetings of advisory committees and conferences we attended, and to others who did likewise in less formal settings, we are grateful.

We owe a special debt to Dr. Howard Myrick, former director of the Office of Communication Research of the Corporation for Public Broadcasting. Not only was Howard instrumental in obtaining financial support from CPB for our undertaking of the present effort, but he subsequently provided us with virtually unlimited freedom in choosing the direction the book would take.

Without the support of The John and Mary R. Markle Foundation, under the presidency of Dr. Lloyd Morrisett and the program management of Jean Firstenberg (currently Director of the American Film Institute), the project would never have been undertaken at all. We are especially grateful for their patience and moral support during the formative stages of its inception.

Michael Rice, Program Director at the Aspen Institute for Humanistic Studies, has been especially helpful in his incisive comments on how we might increase awareness and acceptance of our work among those responsible for PTV programming.

Finally, there were those at National Analysts whose hands-on efforts were indispensable in getting the work done. Lyn

Wiesinger served as project manager from the questionnaire development stage through the creation of final tabulations and multivariate analyses. Tony Asmann, Director of Data Collection Services, and Mary Henderson drew the sample and developed the weighting model for the data. Nancy Lessin, Field Department Manager, and Ethel Trachtenberg supervised the interviewing. Jim Clody, and later Marie Metz, supervised the preparation and processing of data. The many drafts of the manuscript for this book were typed by Nettie Massimi.

We are solely responsible, however, for those errors of judgment or fact that remain, despite the best efforts of those mentioned above and others who tried to help us avoid them.

1

Perspectives

The survey findings reported in this book describe public television from the viewpoint of its audience. They provide new insights into who watches public television and, more importantly, why they watch it. A better understanding of the many ways people use PTV will be useful to anyone with serious interest in the medium whether they are concerned with its role in society, its effect on certain segments of society, what programs to produce or air, when to air them, how to promote them, or how to evaluate existing programs or stations.

In a previous book entitled *The Public's Use of Television* (Frank and Greenberg, 1980) findings were reported from the same survey. However, our emphasis was on understanding television in general, with only a single chapter dedicated to PTV. Because of the importance of PTV in our society, and with the continuing interest and sponsorship of both the Corporation for Public Broadcasting and The John and Mary R. Markle Foundation, we decided that this book should be written.

The pages that follow report a detailed analysis of the patterns of leisure interests and needs of the American audience and how they relate to that audience's use of public television. Toward this end, the audience is subdivided or segmented into fourteen distinct groups, each of which has a relatively unique pattern of leisure interests and needs. Our approach is one way of characterizing the diversity of the American audience in terms that we believe are useful in helping PTV develop and implement strategies to achieve its current programming objectives. Such strategies are particularly important in this period when emerging changes in telecommunications technology promise to revolutionize the television industry. These changes will provide American audiences with a greater number of telecommunications alternatives from which to choose, and will increase the already formidable competition for viewers faced by PTV in the form of the commercial networks. The linkage between these emerging technologies and the goals of PTV is discussed below, followed by an explanation of how our approach to segmenting audiences can contribute to strengthening the role of PTV.

Goals

Public broadcasting, as it is constituted today, evolved from the recommendations of the Carnegie Commission on Educational Television, which were published in 1967. Since its origin there has been basic agreement, at a rather abstract level, as to some of the goals it should seek to accomplish. This has been most eloquently stated by E. B. White in an often-quoted letter written to the Commission:

> Noncommercial television should address itself to the ideal of excellence, not the idea of acceptability—which is what keeps commercial television from climbing the staircase. I think television should be the visual counterpart of the literary essay, should arouse our dreams, satisfy our hunger for beauty, take us on our journeys, enable us to participate in events, present great drama and music, explore the sea and the sky and the woods and the hills. It should be our Lyceum, our Chautaugue, our Minsky's, and our Camelot. It should restate and clarify the social dilemma and the political pickle. Once in a while it does, and you get a quick glimpse of its potential [Carnegie Commission, 1967: 13].

In recommending the establishment of a public television system, the Commission defined in part the criteria that programs to be aired on PTV should meet:

> The programs we conceive to be the essence of Public Television are in general not economic for commercial sponsorship, are not

designed for the classroom and are directed at audiences ranging from tens of thousands to the occasional tens of millions [Carnegie Commission, 1967: 3].

The Commission then went further in defining the essence of what it felt public television could contribute to American society:

> We have come to see that since the technology of television lends itself readily to uses that increase the pressure toward uniformity, there must be created means of resisting that pressure, and of enlisting television in the service of diversity. We recognize that commercial television is obliged for the most part to search for the uniformities within the general public, and to apply its skills to satisfy the uniformities it has found. Somehow we must seek out the diversities as well, and meet them, too, with the full body of skills necessary for their satisfaction.
>
> America is geographically diverse, ethnically diverse, widely diverse in its interests. American society has been proud to be open and pluralistic, repeatedly enriched by the tides of immigration and the flow of social thought. Our varying regions, our varying religious and national and racial groups, our varying needs and social and intellectual interests are the fabric of the American tradition.
>
> Television should serve more fully both the mass audience and the many separate audiences that constitute in their aggregate our American society. There are those who are concerned with matters of local interest. There are those who would wish to look at television for special subject matter, such as new plays, new science, sports not now televised commercially, music, the making of a public servant, and so on almost without limit. There are hundreds of activities people are interested in enjoying, or learning about, or teaching other people. We have been impressed by how much we might have from television that is not now available.
>
> To all audiences should be brought the best energies, the best resources, the best talents—to the audience of fifty million, the audience of ten million, the audience of a few hundred thousand. Until excellence and diversity have been joined, we do not make the best use of our miraculous instrument.
>
> The utilization of a great technology for great purposes, the appeal to excellence in the service of diversity—these finally became the

concepts that gave shape to the work of the Commission [Carnegie Commission, 1967: 13-14].

Almost a decade later, the basic positioning of public broadcasting (not just television) to strive for excellence and respond to diversity pervades the stated mission and goals of the Corporation for Public Broadcasting (1976) as well as the report of the second Carnegie Commission (1979). Both of these documents go somewhat further in defining certain audience-related elements of the goals.

The Corporation for Public Broadcasting (CPB) statement recognizes that there are many American publics and that public broadcasting should strive to meet their diverse needs; that it should strive not only to offer programs that meet the needs and desires of small groups, or classes of people but, collectively, those of all the American people. Furthermore, it states that the CPB should take the lead in the development of effective techniques to ascertain the audience's needs and interests.

The second Carnegie Commission report is a bit more specific. It recommends that public television seek to serve all Americans so that 100% of the potential audience is served on a regular basis. They, in turn, suggest that at first, "serving on a regular basis" be defined as reaching each individual at least once a month and that a long-term goal of at least once a week be established. In 1975, 49% of the potential nationwide audience viewed noncommercial television at least once a month. By 1978 the figure had risen to 63%. In setting out this audience objective the Commission was careful to state that its concern was for the overall cumulative reach of all programs offered, not just that for one program or program type. Consistent with the concept of diversity, an extreme of, say, 100 programs, each of which attracted 1% of the potential audience, and whose viewers did not overlap, would achieve the Commission's goal. Clearly this goal differentiates PTV's audience objective radically from those associated with commercial television, wherein a program attracting 1% of the audience would be a financial disaster.

These statements of goals regarding PTV's role, though different in language, are nonetheless consistent with the Public

Telecommunications Financing Act of 1978 as well as with the prior legislation, the Public Broadcasting Act of 1967. In the more recent of the two, emphasis is placed on the development of both public radio and television, including their use for instructional, educational, and cultural purposes. This effort is combined with the stated goal of making public telecommunication available to all citizens of the United States.

In spite of PTV's goal of serving interests and needs that are idiosyncratic to various diverse segments of American society, until quite recently few research efforts have studied the motivations for public television viewing. The first major published study of television that paid the slightest attention to PTV was conducted by Bower (1973) in which about six of 200 pages are devoted to educational television. The first widely disseminated study devoted exclusively to PTV was reported by Lyle (1975). The Lyle study provided the first road map of both how much PTV was being watched and by whom. Viewers were characterized on the basis of population demographic and socioeconomic characteristics, such as geographic location, education, and stage of life cycle. Neither Lyle nor Bower made a direct effort to analyze audience needs and interests. The study reported in this book was designed to address precisely these audience characteristics.

It is clear that there is a growing interest and commitment on the part of the CPB to obtain a better understanding of the motivational dynamics of PTV viewing behavior. Its recent efforts relating to research strategies for measuring television's audience in qualitative terms (Corporation for Public Broadcasting, 1980) and to the effects of television on people's lives (Corporation for Public Broadcasting, 1978) complement our effort.

Emerging Technologies

Cable television, once limited to a few rural areas, is rapidly expanding not only in rural, but in major metropolitan areas as well. Many organizations are intent on exploiting this change by offering various pay-TV services such as recent movies and/or sports. At the same time, we are witnessing the introduction of videodisk, videotape, and video-interactive equipment. If the rapid market growth of these electronic marvels parallels that of other electronic products, as we think it will, it is reasonable to expect sharp price declines in real dollars followed by even more rapid growth. Meanwhile, even newer technologies, such as fiber optics, may one day substantially increase the number of channels a television set can receive over a cable. These changes will provide television audiences with a broader range of options on their screens, with increased personal controls over what they watch and when they watch it.

As we write this book it is becoming increasingly recognized that the expanded numbers of options available to television audiences are apt to lead not only public television, but also cable television and, to some extent, the commercial networks, toward more diversified programming. It has been demonstrated time and time again in the marketplace that as markets become increasingly fractionated or segmented, the market responds by offering a variety of highly specialized products designed to meet the needs of smaller and smaller market segments. There is no reason to believe that the television industry taken as a whole will respond differently. General recognition of this fact is gradually occurring both in the public press (Bernstein, 1979; Funt, 1979) and in more scholarly tomes (Robinson, 1978).

Will the 1980s Be a Watershed for PTV?

Both because of the long-avowed goals of PTV and the aforementioned technological imperatives, understanding audience interests and needs on a segment-by-segment basis has evolved from an activity that would be interesting to do someday to one that is of central and immediate importance.

Perhaps these technological changes will lead the CPB to face increased competition from commercial organizations that find it economically necessary and desirable to develop programming aimed at interests and needs relatively unique to small segments of society, rather than programming directed only toward mass audiences. Whether or not PTV *chooses* to compete with commercial organizations, viewers may increasingly regard a subset of commercial offerings as competitive with (substitutes for) one or more elements of PTV programming.

PTV's commitment to excellence in the future may be jeopardized not so much by the risk of a lower standard of excellence, but by the fact that other institutions may see their opportunities as, in part, overlapping PTV's (even though they may or may not lack the same commitment to excellence). Those institutions may do as good, or even a better job, of understanding the diversity within American society and of tailoring their program offerings to it. Our purpose in advancing this line of argument is neither to praise nor condemn PTV, but to offer an analysis of what is likely to be the emerging competitive framework in which PTV will be attempting to achieve its objectives.

Findings such as those reported in this volume can contribute to the efforts of those reaching for new highs in PTV's attempts to deliver a standard of excellence in programming that is meaningful to the diverse segments that constitute the American television audience.

Study Overview[1]

In developing our approach to understanding the uses of both public and commercial television we have drawn from experience reported in two streams of literature:

(1) *market segmentation*—appearing almost completely in the field of marketing; and
(2) *needs and gratifications*—appearing in the fields of sociology and communication.

It is not our purpose to review this material as this has been done elsewhere (Frank, Massy, and Wind, 1972; Wells, 1974; Blumler and Katz, 1974).

The literature of market segmentation is in large part supplemented by our personal experiences gained in conducting numerous proprietary studies of consumers across a wide range of products and services, for commercial firms marketing products and services and for their advertising agencies. This experience supports our contention that one of the most useful ways to study consumer behavior is to obtain data, not only on consumers' product usage and demographics, but also on their attitudes, interests, and opinions (AIOs) specific to the product or service category under investigation. For many food products researchers have been led to study consumers' attitudes toward meals and meal preparation. For example, knowing how confident a cook is in preparing meals, in trying new foods when company comes, or how important family nutrition is, is extremely helpful in understanding why new convenience food products are more readily accepted by some consumers than are others. Such studies have allowed marketers to develop successful new products and to direct their marketing efforts most efficiently against those target consumer segments likely to be most receptive to such products.

Based on exploratory research (National Analysts, 1975) we conducted prior to our national study, we concluded that the most relevant AIO context for the study of both public and commercial television would include measures of a wide range of leisure interests and activities (e.g., football, opera) and subjects (e.g., international affairs, astronomy, poetry). For many people, leisure-related activities are the principal complements or substitutes for watching television. Another class of potential complements or substitutes is other media such as movies, radio, newspapers, books, and magazines. Measures of the usage of these media are included in the present investigation. Though we are far from the first to recognize the need to study television in general and public television in particular in this context (Morriset, 1976), we believe we are the first to attempt to extensively investigate their relationships.

It is also apparent from the literature on needs and gratifications that, to understand the uses of a product or service from the consumer's point of view, it is useful to obtain data on the psychological needs satisfied by its use. Hence, we have chosen to develop and include a set of need-related questions about which more will be said in the next two chapters.

It is helpful to view the elements of our approach, as discussed in the preceding paragraphs, in terms of attempts to answer the following six questions:

(1) What are the patterns of leisure interests and activities that characterize different segments (groups) of the American public? How do they differ in the psychological needs gratified by their leisure interests and activities? How do they differ in their demographic characteristics?

(2) Which segments are heavy PTV viewers? Which are light? Why? What are the nature and extent of corresponding differences with respect to each segment's usage of commercial television? Books? Magazines? Movies? Newspapers? Radio?

(3) What are the extent and nature of heavy- and light-viewing segment differences by PTV programs? By the types of programs people wish to see more of on PTV? By new program concepts?

(4) How do voluntary giving and general funding preferences for PTV vary as a function of different segments' PTV viewing behavior?

(5) How do Black, Hispanic, and elderly minorities differ from the general population in their use of PTV and other media?

(6) What programming strategies might serve to attract persons in each segment? How are these strategies apt to differ across segments?

Our findings and conclusions are based on a representative nationwide survey of 2476 people aged 13 and over, supplemented by a special sample of 278 Hispanics. These people were interviewed in person for an average of an hour and one-half each. Besides obtaining an extensive description of their television viewing behavior, the interview included questions pertaining to interests in leisure activities and various subject matter, the needs served by these interests and their use of other media.[2]

The data on interests were used to assign individuals into segments such that the members of each segment have a relatively homogeneous pattern of interests, while differences across segments are great. For example, the people in one segment have strong interests in mechanical activities and outdoor life, while those in another are predominantly interested in artistic and cultural activities. Examining the differences in television and other media usage across these interest segments leads to an increased understanding of the determinants of PTV viewing behavior and provides a valuable conceptual framework for the development and implementation of strategies for audience attraction.

Relationship to Previous Research

The Work of Others

In addition to the work of Bower (1973) and Lyle (1975), we found only two other studies that are similar in spirit, if not in exact detail, to ours, namely: *The People Look at Television* (Steiner, 1963), and *On the Use of Media for Important Things* (Katz, Gurevitch, and Haas, 1973).[3]

Steiner's book is the predecessor to Bower's, and is based on a nationwide sample of the U.S. audience. The Bower book replicates and updates Steiner's work. Both studies provide detailed descriptions of audience demographic and socioeconomic characteristics, but examine only to a modest extent the needs served by television (e.g., entertainment, education). The second study (Katz et al., 1973) is of interest primarily because the authors attempted to develop an overall structure of the functions performed by the mass media including, but not restricted to, television.

The study reported in this volume goes beyond these investigations principally in three ways:

(1) It focuses on the use of PTV, not as an isolated phenomenon, but in terms of its relationships with a broad range of interests in leisure activities and subject matter.
(2) It focuses on the uses of PTV in relation to uses of other media, again as opposed to studying it in isolation.
(3) It develops a segmentation scheme to describe how different types of audience members with varying interest patterns use PTV.

We decided to broaden the scope of our inquiry to include a wide range of leisure interests, as well as other media, for the following reasons:

(1) Leisure interests and other media comprise the principal alternatives to television usage. This is reasonably well recognized in the

literature on the subject of leisure (e.g., Kaplan, 1975), but is often ignored in studies primarily concerned with television.
(2) An exploratory study consisting of a series of group depth interviews (National Analysts, 1975), and the pilot survey we conducted prior to the national survey (Frank and Greenberg, 1976), strongly supported the usefulness of including both interests and other media for the purpose of better understanding television usage.
(3) Our combined past experience in designing and conducting dozens of large-scale segmentation studies of this general type for a variety of products and services supports the utility of examining the product or service that is the focus of the study in a broader context.

We have used a *people types* analysis rather than a *variables* analysis approach in the study. It is our intent to gain an understanding, to the extent possible, of the totality of a person's relationship to PTV as opposed to intensively studying some limited aspect of a person (say, income) and its relation to viewing behavior. Focusing on understanding people in terms of attitudinal and behavioral patterns rather than in terms of individual variables leads one in two fundamentally different directions from the aforementioned studies, namely:

(1) grouping people based on their patterns of interests rather than on their responses to a single interest; and
(2) interpreting results in terms of each segment's profile of answers across all variables rather than examining them one variable at a time. For example, we find it more useful to know that a person is generally interested in cultural activities than to know of an interest in any single activity.

One last comment on the relationship of our research approach to the three previously mentioned studies is in order. Though the present study examines interests and needs in considerably more detail than do the others, it has in common with them the fact that the measures included are situationally dependent. That is, they were developed for the express purpose of segmenting media audiences based on interests and hence, may have a clear-cut contextual relation to that subject matter. In contrast, we could have proceeded to use much more generalized scales such as those found in standard personality tests. Our use of situationally dependent measures is consistent with the general direction taken by proponents of segmentation research in marketing to increase the chances that their findings will be both interpretable and useful in developing effective marketing strategies (Wells, 1974).

Our Own Research and Development Efforts

Two exploratory studies were conducted prior to the project reported in this book. Their purpose was to contribute to developing and evaluating the methodology to be used in the national study.

The first of these was an exploratory study conducted primarily to generate hypotheses as to what needs are important determinants of leisure interests, especially as they relate to television viewing behavior (National Analysts, 1975). For this purpose, sixteen group depth interviews were conducted in varying geographic locations with between six and eight participants in each. The geographic locations were Philadelphia, Chicago, Los Angeles, Nashville, and Wilkes-Barre, Pennsylvania. The findings of this study helped to identify the needs to be measured and to define the language to be used in asking about them. This preliminary project was completed in April 1975.

It was followed by a demonstration project (Frank and Greenberg, 1976) in which the survey procedure under consideration at the time was "dry run" on a sample of 1200 respondents. The purposes of this project were:

(1) to check the technical aspects of the survey such as the length of the interview, the willingness of respondents to cooperate, questioning sequence, and wording; and
(2) to provide actual data using the intended analytic approach to demonstrate, in considerable detail, the type of results to be expected from a national study. Had it not been for this objective a considerably smaller sample size would have been adequate.

This latter objective was necessary to address the concerns expressed by members of an advisory committee that reviewed the results of the exploratory study and our proposal for a national survey. Questions arose as to whether leisure interests and associated needs would be related to viewing behavior at all, let alone in a way that might be useful to the television industry.

The demonstration project was successfully completed in November 1976.

Plan of the Book

The next chapter describes the design of the study, followed by Chapter 3, which introduces each of the segments based on their interests, needs, and demographic characteristics. Chapter 4 reports the findings on PTV usage. Chapters 5 and 6 discuss PTV's relationship to other media, namely, commercial television, books, magazines, movies, newspapers, and radio. Chapter 7 discusses PTV program preferences. This is followed by a report in Chapter 8 of the PTV voluntary donation practices of the American audience as well as viewers' preferences for alternative types of funding. Chapter 9 discusses the PTV viewing behavior of Blacks, Hispanics, and the elderly. Finally, Chapter 10 discusses the implications of our findings for PTV.

Notes

1. Much of the material in this section appears in Frank and Greenberg (1980). Where necessary it has been appropriately modified, given the objective of this book.

2. Chapter 2 provides a more detailed description of both the sampling procedures used and the types of data collected.

3. Those interested in a comprehensive review of television literature should see Comstock (1975), Comstock and Fisher (1975), Comstock and Lindsey (1975), Katz (1977), and Comstock, Chaffee, Katzman, McCombs, and Roberts (1978).

2

Study Design

The data for this study were obtained from a national survey of 2752 television households. The first section of this chapter describes the data collection process—both the content of the questionnaire and the sampling and extrapolation procedures used for developing population estimates. The survey population consists of people aged 13 and over residing in U.S. households with at least one television set. The second section reports the analytical strategy that serves as the principal basis for the findings reported in subsequent chapters.

Data Collection Procedures

Questionnaire Content

The following is a brief description of the questionnaire's contents.[1] As the results are discussed in subsequent chapters, and where necessary, a more detailed description is provided. The major categories of coverage are as follows:

(1) for 139 *leisure interests and activities,* the degree of interest respondents had as measured by their response to a 4-point scale;
(2) for 59 items characterizing *needs satisfied by their leisure interests* (e.g., "to kill time," "to feel I am important to other people"), their importance as measured by a 4-point scale;
(3) for 150 specific television programs (22 of which were aired on PTV), *recall measurements* of:

   (a) program viewing behavior
   (b) decision-making behavior—Who made the decision to watch?

(c) viewing context—Who else watched?
   (d) viewing importance—Is it important not to miss? Does it serve as a background for doing other things?
(4) *additional TV-related behavior,* including overall estimates of the viewing of public and commercial television, as well as information relating to sources of information about programming and reasons for viewing;
(5) a number of questions related to *selected aspects of PTV,* namely:
   (a) the image of both public and commercial television,
   (b) types of programs viewers would like to see on PTV,
   (c) ratings of new program concepts developed for PTV, and
   (d) attitudes toward various sources of funding for public television, as well as past contributions to local PTV stations;
(6) *media usage for books, magazines, movies, newspapers, and radio;*
(7) *television set ownership, commercial and public broadcasting coverage and awareness;* and
(8) *demographic and socioeconomic characteristics.*

Sample Design

Two different sampling procedures were used to select the 2752 respondents. The principal sample of personal interviews with 2476 persons was based on a national area probability sample drawn to permit the projection of results to the entire population of the conterminous United States aged 13 and over living in households with one or more television sets. In addition, a supplementary sample of 276 Hispanics was selected using quota sampling procedures in areas of high Hispanic concentration.

National Probability Sample—2476 people. For the selection of this sample conventional area probability sampling procedures designed to provide a rigorous statistical basis for projecting findings to the entire population were employed. The sampling process provided strict controls over the selection of respondents, starting with the selection of large geographic areas via a series of steps going all the way to the selection of small area segments, households, and individuals within households.[2] Blacks were disproportionately sampled to ensure a large enough sample size

STUDY DESIGN

for the analysis of minority audiences (Chapter 9).[3] A statistical weighting procedure in the analysis stage brought Blacks back into balance with their incidence in the population.

The selection process included making up to three callbacks to contact eligible respondents in each household. Fieldwork was monitored by National Analysts' staff, who checked and edited questions promptly so that any ambiguities could be resolved by contacting the interviewer. Telephone validations were conducted with 15% of the respondent households.

The interviewing was conducted during the period beginning October 15, 1977, and ending January 7, 1978. The interview was unusually lengthy, averaging about 90 minutes.

The data from the questionnaire were keypunched and subjected to 100% keypunch verification. A mechanical edit to check for the internal consistency of each interview was conducted and problems were resolved by National Analysts' staff. The data from this sample form the basis for all of the results discussed in this book except for those relating to Hispanics, which are based upon the supplementary sample of 276 Hispanics described below.

A sample weighting procedure was used to develop estimates for the population aged 13 and older. The findings reported based on the national probability sample of 2476 persons, are estimates for the entire U.S. population and not simply raw counts taken from the original sample.

Hispanic Sample—276 people. The "supplementary" Hispanic sample is not included in the data used to develop the estimates for the entire population since the questionnaire content and the sampling procedures used for this group of persons were not entirely comparable. The questionnaire was translated into Spanish for those who were not bilingual and hence it is difficult to be sure that the questions contained in it have precisely the same meaning, although an attempt was made to do this. Despite these efforts, the type of Spanish spoken and understood varies across different regions of the U.S. depending on the origins of the local Hispanic population. In addition, we were forced to use a quota sampling procedure of this segment in a number of locations (15) in which relatively high concentrations of Hispanics are located. They are:

Albuquerque, NM	Lea, NM
Bronx, NY	Los Angeles, CA
Chicago, IL	Miami, FL
Houston, TX	New York, NY
Jersey City, NJ	Phoenix, AZ
Kern County, CA	Queens, NY
Kings, NY	San Diego, CA
San Francisco, CA	

Within each of these areas interviewers were given quotas specifying the number of Hispanic adults and children, aged 13 and over, residing in households with at least one television set, to be interviewed.

This supplementary sample of 276 Hispanics is used as the data base only in the chapter on minorities. There are an additional 69 Hispanics who were selected and included as part of the national probability sample. These respondents were bilingual and were interviewed in English. For the chapter on minority audiences we felt we needed both a larger sample of Hispanics and one that included people who did not necessarily speak or understand English.

Analytical Strategy

A major purpose in conducting this study was to gain insight into people's use of PTV by examining their interests in activities and various subject matter, as well as the needs that these interests fulfill. Given this purpose the question arises as to what interests should be measured. Should sports be included? What about homemaking-related activities, such as food preparation or serving? What needs should be measured? Should we include the need to spend time with friends? To kill time? To learn new thoughts or ideas?

Television viewing remains an extremely flexible means of individual expression. It can help satisfy an almost endless array of needs in relationship to an almost equally endless array of

STUDY DESIGN 35

interests. This flexibility presents a difficult problem to those who wish to analyze, in quantitative terms, the interests and needs that affect viewing behavior. Because of the ubiquity of television one can make an argument for including virtually any interest or need.

The following sections discuss the interests and needs included in the questionnaire, the reasoning that led to their inclusion, and the "summary" measures, based on responses to the aforementioned interest and need questions, that were actually used to study television and other media behavior. Also discussed is the process by which the summary measures of individual interests were used to group individuals into categories (segments) of people with varying patterns of interests. The segments resulting from this last process are introduced in Chapter 3, and their television and other media habits are analyzed in detail in Chapters 4 through 9.

Interests

Measurement Strategy. All those participating in the study rated their interests in each of 139 activities and subjects. This set of 139 interests covers an extremely broad range of activities, including:

(1) *active versus passive*—e.g., camping, listening to radio
(2) *individual versus group*—e.g., auto repair, community social functions
(3) *home versus nonhome*—e.g., meal preparation, fishing
(4) *popular culture versus special interest*—e.g., visiting friends, opera

Also covered is a broad range of subjects such as politics, science, human behavior, and transportation.

In selecting the final set of 139 items from a virtually infinite number of possibilities, we tried to be comprehensive in our coverage of content areas that might be related to television viewing behavior, while at the same time avoiding the level of specificity and detail that might be more appropriate in defining a target audience for a particular program. Our objective was to develop an interest segmentation scheme that would be sufficiently generalizable to be useful in the context of programming decisions that cut across a wide variety of content areas, including news, sports, science, drama, soap operas, and many others. Given this objective we felt it necessary to provide a breadth of coverage with a relatively small number of items representing each content area examined, recognizing that this would limit our ability to explore any single content area in depth. For each of the 139 items, respondents were asked to indicate their degree of interest on a 4-point scale: 1 (not interested at all); 2 (not very interested); 3 (quite interested); 4 (extremely interested).

Use of Summary Measures. Though the 139 interest ratings provided the interest category coverage that was desired, they nonetheless posed two problems regarding their use in the remainder of the analysis. Working with such a large number of items renders the interpretation of results difficult and the data processing burdensome. In addition, given our purpose of developing a "generalizable" interest segmentation scheme, it seemed more appropriate to work with some reduced set of summary measures.

A statistical method referred to as Principal Components Analysis was used to operationally define a set of summary measures (called factor scores) for use in the remainder of the study. In effect, this analysis assisted us in measuring simultaneously the actual pattern of answers to our 139 interest questions and in using the results as a basis for combining questions into a set of summary measures. Rather than use our a priori judgment as to what categories to use, we used data taken directly from the respondents.[4] Figure 2.1 lists the labels we chose for each summary measure (hereafter called factor) in our subsequent analysis, together with an illustrative list of interests that comprise the factor and contribute to each respondent's score on the factor. This statistical procedure permitted the reduction of the 139 original measures to a set of 18 factor scores for each respondent. Each of these 18 factor scores is a weighted average of the original 139 interest scores. Thus, an individual would receive a high score on a factor by expressing extreme interest in the activities or subjects associated with it and would receive a low

Figure 2.1
Individual Interest Factors

(1) Comprehensive News and Information
- national economy
- unemployment
- tax laws
- legal process in U.S. courts
- state issues
- morality in politics
- social security system
- preventive medicine

(2) Athletic Activities—Participant
- snow skiing
- water skiing
- tennis
- volleyball

(3) Household Activities and Management
- housecleaning
- meal preparation
- sewing
- needlework
- household management

(4) Classical Arts and Cultural Activities
- opera
- classical music
- ballet
- live theater
- literature
- painting

(5) Reaping Nature's Benefits
- fishing
- hunting
- agriculture and farming
- gardening

(6) Professional Sports
- baseball
- basketball
- football
- boxing
- golf
- hockey

(7) Science and Engineering
- chemistry
- electronics
- medical sciences
- engineering
- geology

(8) Popular Entertainment
- visiting friends
- radio
- travel/sightseeing
- popular music
- dining out

(9) Religion
- religious organization activities and religion

(continued)

Figure 2.1 Continued

(10) Popular Social Issues
- sex education
- sexual attitudes and behavior
- rights of minority groups
- occult
- women's rights

(11) Indoor Games
- board games
- crossword/jigsaw puzzles
- chess/checkers
- playing cards

(12) Community Activities
- community social functions
- charities and civic associations
- local cultural activities

(13) Investments
- real estate investment
- managing a business
- stock market

(14) International Affairs
- arms race
- balance of trade
- conflict in the Middle East

(15) Camping Out
- camping
- backpacking
- hiking
- boating

(16) Crime and Society
- capital punishment of criminals
- abortion vs. right to life issue
- causes and prevention of crime

(17) Mechanical Activities
- auto repair
- auto racing
- model building
- engineering
- electronics

(18) Contemporary Dancing
- modern dance
- dancing

---

score by expressing little or no interest in those activities or subjects.

Needs

A battery of 59 need questions was constructed to help understand the motivations behind the pursuit of the various leisure interests and activities described above. The items themselves were drawn from a wide variety of sources, including the earlier group depth interview phase of the present project and selected items from other studies. The battery was modified after the completion of the pilot study to eliminate items that were highly redundant and to include additional items where they appeared necessary to flesh out our understanding of the dynamics of leisure interest patterns.

The earlier study consisted of a series of group depth interviews conducted among individuals with a diversity of age and sex characteristics (National Analysts, 1975). A major purpose of that study was to develop a conceptual framework and a set of working hypotheses for understanding the relationships among patterns of leisure interests, the needs addressed by them, and television behavior. That study served as the principal basis for developing the 59 items that were used.

The need-satisfaction items were written to ensure that they dealt with knowledge, action, and feelings as well as:

(1) *basic maintenance*—needs associated with a person's minimum requirements for normal existence; needs that serve to protect one from a state of extreme psychological deprivation or from pronounced psychological stimulation;
(2) *social*—needs related to one's interactions with other people; and
(3) *self-actualization*—needs related to self-development and growth, maturity, and the broadening of one's horizons.

The literature on television, as well as that associated with leisure interests, was of little help in developing the battery of 59 items. The most relevant article was one by Katz, Gurevitch, and Haas (1973), which reported the results of a study on the use of mass media.

Figure 2.2
Individual Need Factors

(1) Socially Stimulating
- to find that my ideas are often shared by others
- to be interesting and stimulating to other people
- to do things which I am familiar with
- to feel good about life in general

(2) Status Enhancement
- to impress people
- to feel more important than I really am
- to have more influence on other people
- to be like other people
- to compete against others

(3) Unique/Creative Accomplishment
- to really excel in some area of my life
- to be more of a leader
- to feel unique and different from other people
- to feel creative

(4) Escape from Problems
- to get away from the pressures and responsibilities of my home life
- to get away from pressures of work
- to relax
- to forget my problems for a while

(5) Family Ties
- to feel closer to my family
- to spend time with my family
- to develop strong family ties

(6) Understanding Others
- to better understand how other people think
- to better understand why people behave the way they do

(7) Greater Self-Acceptance
- to lift my spirits
- to understand myself better
- to overcome loneliness
- to feel I am using my time in the best way possible

(continued)

Figure 2.2 Continued

(8) Escape from Boredom
- to be entertained
- to kill time
- to escape from the reality of everyday life
- to experience again events and places I enjoyed in the past

(9) Intellectual Stimulation and Growth
- to find out more about how things work
- to learn new thoughts and ideas
- to learn about new things to do
- to learn about new places to see

Respondents were asked to rate the importance of each of the 59 need items as reasons for their degree of interest in the leisure interests and activities battery they had just completed. Each of the needs was rated on a 4-point scale: 1 (not at all important); 2 (not very important); 3 (quite important); 4 (extremely important).

These 59 need items were analyzed using the same statistical procedure as was applied to the interest items. This led to the use of nine factors as the basis for computing our summary measures. These are summarized in Figure 2.2.

Because the interpretation and labeling of these factors is less straightforward than for the interest factors, a brief discussion of each is provided below.[5]

(1) *Socially Stimulating.* Persons scoring high on this factor appear to have an above average need to interact with others and to be seen as interesting and stimulating by them. Note that the need expressed here is oriented toward being the kind of person that is stimulating to others rather than toward receiving social stimulation.

(2) *Status Enhancement.* The need expressed in this factor is to gain self-respect and self-confidence by impressing others and influencing them. Presumably those scoring high on this factor pursue their leisure interests and activities with an eye toward how they will be perceived by those people they seek to impress.

(3) *Unique/Creative Accomplishment.* This factor appears to reflect a need to "pull away from the pack" through the achievement of

excellence in some domain. The items focus on a need to strengthen one's sense of individuality and identity by engaging in and succeeding in creative activities. It differs from the previous one in that the present need satisfaction appears to be less dependent upon the perceptions of others and more upon the individual's feelings about him/herself.

(4) *Escape from Problems.* This one is relatively self-explanatory. People scoring high are using their leisure interests and activities to relax and get away from the stresses and responsibilities of everyday life at home and at work.

(5) *Family Ties.* All of the items important to this factor concern the need for strengthening the bonds with one's family.

(6) *Understanding Others.* People scoring high on this factor appear to be seeking greater insight into the thought processes and the behavior patterns of others, possibly as a vehicle to better self-understanding.

(7) *Greater Self-Acceptance.* The need expressed here seems to be one of mood elevation. High scorers presumably view their leisure interests and activities as a means of lifting their spirits and enhancing their feelings of self-worth and self-acceptance.

(8) *Escape from Boredom.* Unlike escape from problems (described above), the need for escape here focuses on relief from boredom rather than from the problems of work and home life.

(9) *Intellectual Stimulation and Growth.* The pattern associated with this factor clearly reflects the need for continued learning and growth. One gets a sense of intense curiosity and concern for a continuing pattern of self-development among those who score high on this measure.

The Interest Segmentation Process

The objective of the interest segmentation process we used was to classify the 2476 respondents in the sample into subgroups that would be relatively homogeneous with respect to their patterns of leisure interests and activities. This was done using a statistical clustering procedure developed by Howard and Harris (1966).[6]

This procedure enabled us to classify respondents into different groups (hereafter called segments), wherein the people in each segment have patterns of interest scores (based on all eighteen of their interests) relatively similar to others assigned to the same segment, and relatively dissimilar from those people assigned to other segments. Each person was assigned to one and only one segment.

STUDY DESIGN

The output of this process, together with an extensive analysis of the results, led us to a decision to classify respondents into fourteen segments—each with its own unique mix of interests. Both this analysis and the analyses of interests and needs were conducted only on the national probability sample of 2476 respondents. Once these respondents had been grouped into 14 segments, we classified the supplementary sample of 276 Hispanics into the same basic 14 segments.

And now for the results. . .

Notes

1. A complete copy of the questionnaire is published elsewhere (Frank and Greenberg, 1980: app. A).

2. For a detailed description of the sampling procedures used, see Frank and Greenberg (1980: ch. 2, app. B and C).

3. Funds for interviewing this additional sample of Blacks as well as the supplementary sample of Hispanics were generously provided by the Corporation for Public Broadcasting.

4. For a complete listing of the interests rated, as well as the details of the factor analysis of the interest scores, see Frank and Greenberg (1980: ch. 3, app. D).

5. For a complete listing of the needs rated, as well as the details of the factor analysis of the need scores, see Frank and Greenberg (1980: ch. 3, app. E).

6. For a more detailed description of the Howard-Harris technique and how it was used, see Frank and Greenberg (1980: ch. 3).

3

The Audience Interest Segmentation

In this chapter we introduce a classification system for the U.S. population of television viewers, based on their patterns of leisure interests and associated psychological needs. The population is divided into fourteen interest segments or types emerging from a cluster analysis of the interest factor scores described in the concluding pages of the previous chapter.[1]

The reader will meet fourteen new friends each of whom has, to some extent, a differentiated set of interests, needs, and demographic and socioeconomic characteristics. The introduction to these friends is accomplished in two stages. First, a brief thumbnail sketch of each segment is presented to provide an overview of its members before becoming immersed in a detailed description of their characteristics. This is followed by a section containing a more detailed analysis of the people in each segment. This chapter presents our general conclusions about the segments, followed by a detailed discussion of the supporting data.

It is especially important for the reader to thoroughly understand the nature of each of the fourteen segments, as subsequent chapters focus on their PTV viewing and related behavior.

Introducing the Interest Segments

Table 3.1 provides labels for each of the fourteen segments together with data on segment size, age, and sex characteristics.

Table 3.1 Overview of Fourteen Interest Segments

	Population Percentage	Average Age (in years)	Percentage of Females in each Segment	Percentage of Females in Total Population
Adult Male Cencentration				
Mechanics and Outdoor Life	8	29	4	1
Money and Nature's Products	6	53	23	3
Family- and Community-Centered	6	47	17	2
Adult Female Concentration				
Elderly Concerns	8	61	71	11
Arts and Cultural Activities	9	44	69	11
Home- and Community-Centered	8	44	84	12
Family-Integrated Activities	10	35	87	16
Youth Concentration				
Competitive Sports and Science/Enginnering	7	22	5	1
Athletic and Social Activities	4	19	83	7
Indoor Games and Social Activities	4	22	91	7
Mixed				
News and Information	5	47	43	4
Detached	9	46	47	8
Cosmopolitan Self-Enrichment	8	36	59	9
Highly Diversified	8	34	51	8
Entire Population	100	40	52	100

In order to better understand the nature of the people in each segment, it is helpful to organize them in terms of their age and sex composition; hence they are classified into four "supracategories," namely: Adult Male Concentration, Adult Female Concentration, Youth Concentration, and Mixed. First, the supracategories are defined in terms of their age and sex composition and then the individual segments within them are discussed. Unless indicated to the contrary, the data reported in Table 3.1, and in all further tables in the book, are statistically weighted to represent the U.S. population (48 conterminous states) of persons 13 years of age and older.

The fourteen segments are grouped into these particular four supracategories as they contain people who have quite different sex and/or age characteristics. This is true despite the fact that only interest data were used as the basis for creating the segments.

As shown in Table 3.1, the Adult Male Concentration category consists of three segments whose members' average age ranges from 29 to 53 years and whose populations are from 77% to 96% male. The four Adult Female Concentration segments, in contrast, are from 69% to 87% female, with average age ranging from 35 to 61 years. The last supracategory in the table, Mixed, also consists of four adult segments (average age ranging from 34 to 47 years). However, their sex composition is nowhere near as extremely skewed toward either sex as the other two adult categories. This leaves only one other category in the table, namely Youth Concentration. The average age of these segments' members ranges from 19 to 22 years, much lower than those for the other segments. The three segments within the Youth Concentration category do, however, differ with respect to their sex composition. One of them is predominantly male (95%) while the other two are predominantly female (83% and 91%).

Segment Sketches

Adult Male Concentration

The three segments in the Adult Male Concentration category are labeled Mechanics and Outdoor Life, Money and Nature's Products, and Family- and Community-Centered. These labels were chosen with the objective of connoting to the reader the general character of the interests and activities of the people who comprise them. The same is true of the labels used for the remaining eleven segments.

People in the *Mechanics and Outdoor Life* segment tend to be young, adult, blue-collar males whose interests focus on noncompetitive activities emphasizing personal physical accomplishment such as auto repair, fishing, and camping. These are interests that, in turn, do not place substantial requirements on interpersonal cooperation or support. Interpersonal relations are not a primary component of either their interests or their needs. They score well above average on their needs to escape and for unique, creative accomplishment, and their interests appear to provide a vehicle to satisfy these needs.

The members of the *Money and Nature's Products* segment are older males with a somewhat above-average representation of rural retirees. Their interests are related to activities that provide some form of tangible return or product such as fishing, hunting, or investments. They are less interested in active, physical activities such as camping out and professional sports and in those that are culturally upscale or abstract, such as classical arts and international affairs. These are somewhat complacent people who, nonetheless, do feel a need for interpersonal contact and support, especially from their families.

The last segment in the Adult Male Concentration category, the *Family- and Community-Centered* segment, consists of people whose interests include many of those of the Money and Nature's Products segment, but incorporate a broader range of activities related to home and community. These people have a much stronger need for family ties than do those in either of the other two segments. They include a mixture of blue- and white-collar employees with a relatively large percentage living in nonmetropolitan areas.

Adult Female Concentration

As one would expect, the segments in this category have quite different interest profiles from those that have just been described. There are four segments, namely: Elderly Concerns, Arts and Cultural Activities, Home- and Community-Centered, and Family-Integrated Activities.

The members of the *Elderly Concerns* segment are older than those in any of the other segments. A much higher proportion are retirees, widowed, and without children. They have relatively few interests. Those areas in which they are interested, namely religion and news and information, appear to help them maintain a sense of social integration and belonging in the absence of very much direct interpersonal contact. Their needs to overcome loneliness and lift their spirits are quite high. They report relatively little need for creative outlets or for intellectual stimulation.

In sharp contrast are the people comprising the *Arts and Cultural Activities* segment. They are, for the most part, highly educated women who either are themselves, or are married to, a household heads that are managers or professionals. They are interested in a broad range of intellectually upscale activities and subjects, especially the classical arts. They have little need to escape nor are they particularly concerned with improving their peer group status. They do report strong needs for intellectual stimulation and growth and for understanding others.

As the title suggests, people in the *Home- and Community-Centered* segment, tend to be married, adult female homemakers. Their interests are associated almost exclusively with home and community activities. Their greatest needs are for maintaining family ties and being socially stimulating. They report relatively low needs for creative accomplishment or for intellectual stimulation.

The last of the four Adult Female Concentration segments is called *Family-Integrated Activities*. This segment has the highest proportion of adult women with young children. They are interested in a broad range of home and family activities (e.g., indoor games as well as home maintenance). In their homes the presence of young children appears to have an important influence on adult interests. People in this segment score high on the need for maintaining family ties.

It is important to note that, while these segments contain a substantial majority of females, they are far from the only segments in which women are found. In fact, only half the women in the total population fall into these four segments. Fully 29% of all women are in the Mixed category, with the remainder in the Youth segments (14%) and in the Adult Male segments (6%).

Males and teenagers also are quite dispersed across the fourteen segments. At this stage of our discussion it is important to note these findings. A potential misinterpretation is to assume that all women fall in the Adult Female Concentration category. They do not. We are not suggesting that all women can be put into four segments and are like all other women when it comes to their interests. Nor should all men be placed in three categories with only other men. On the contrary, the interests of males and females of all ages tend to be overlapping and hence members of any major demographic group are found in virtually every one of the fourteen interest segments.

What are the implications of this overlap? Though there is an association between these demographic characteristics and segment membership, it is a far from perfect relationship. The interest segmentation complements the demographic segmentation scheme that is traditionally used for studying audiences for television and other media.[2]

Youth Concentration

The three segments in this category are labeled as follows: Competitive Sports and Science/Engineering, Athletic and Social Activities, and Indoor Games and Social Activities. The first of these three contains large numbers of teenage males, while the latter two include many teenage females.

Members of the *Competitive Sports and Science/Engineering* segment are interested in mechanical activities such as auto repair, as well as in competitive sports, both participant and professional. They score quite high on the need to escape from boredom and relatively low on the needs for understanding others and achieving greater self-acceptance.

The females in the *Athletic and Social Activities* segment are from high-income families. Their interests are oriented toward active, away-from-home activities, such as participant athletics and popular entertainment. They have above average needs to be socially stimulating and the highest need to escape from problems along with the lowest need for family ties.

In contrast, the women in the *Indoor Games and Social Activities* segment come from lower-income families. Somewhat older than the previous segment, their interests and needs are more home- and family-related as they have "cut the apron strings," and many are in the process of establishing their own households. They report a high need for status enhancement, but are quite low in their need for unique/creative accomplishments.

Mixed

Each of the four segments in this category is composed of a more balanced mixture of male and female adults. The segments are labeled: News and Information, Detached, Cosmopolitan Self-Enrichment, and Highly Diversified.

News and Information segment members tend to be physically passive adults whose interests center around the collection and dissemination of information on a wide range of subjects and issues. Their needs are focused on family ties and on being socially stimulating.

The *Detached* segment members are characterized by their extremely low levels of interest across each of the eighteen interest factors. These people have a downscale socioeconomic profile. They score relatively low on all of the need factors as well. The leisure interests and activities studied in this project are probably of little relevance in their struggle for economic and social survival.

In contrast, people in the *Cosmopolitan Self-Enrichment* segment are upscale in their demographic and socioeconomic characteristics. They report broad interests spanning intellectual and cultural activities and subjects, and are physically active. They are high on the needs for intellectual stimulation, unique/creative accomplishment, and understanding others. They are especially low in their needs for status enhancement and for escape from boredom.

The last of the fourteen segments is entitled *Highly Diversified*. They are disproportionately southern, Black adults in homes with children. They report very broad interests, especially those relating to activities involving personal participation with family and other informal small group settings. Their strongest need appears to be for intellectual stimulation and growth.

This is the reader's initial introduction to each of these fourteen friends. We hope some of them are recognizable in that they represent one way of characterizing the American population of which most every reader of this book is both a member and an observer.

In the following section each of these segments is discussed once again. This time, however, the discussion of their interests, needs, demographic, and socioeconomic characteristics is more detailed in an effort to develop an in-depth profile of each of them.

The Interest Segments in Detail

In the following discussion a detailed description of the membership of each segment is reported. The data upon which this section is based are contained in Tables 3.2 to 3.5, which report for each interest segment the interest factor scores and the need factor scores along with demographic and socioeconomic characteristics. The discussion centers on the *membership of each segment,* not on *individual variables* such as income. Our purpose is to paint fourteen portraits, in words and numbers. This is quite different from focusing on variables one at a time, for the purpose of evaluating their relationship to interests. The emphasis is on the description of people, not variables.[3] Hence, the discussion of each segment covers the relevant data from all of the tables at once.

In the discussion that follows we have chosen to deemphasize the citing of specific tables and values within them. The tables, however, have been retained in the body of the text so that the reader who wishes to check our reasoning or to determine the specific numerical values that serve as the basis for statements that are made may do so with relative ease.

Each of the measures for which findings are reported in Tables 3.2 to 3.5 has been subjected to statistical testing (using the univariate F ratio) to determine the likelihood that the overall variation in scores across all fourteen segments is due to chance fluctuations. With only one exception, all of the F ratios for all of the variables in all four tables are significant at the .01 level (i.e., if there were, in fact, no differences, there is less than 1 chance in 100 that segment differences as large as we found would have occurred by chance). Most of the differences are significant at the

.005 level (i.e., fewer than 5 chances in 1000). These results for needs, Table 3.3, and for demographic and socioeconomic characteristics, Tables 3.4 and 3.5, respectively, confirm that the general pattern of differences across segments is "real" and not simply the result of chance.

However, the F ratios for Table 3.2 need to be taken "with a grain of salt." They are, at best, just one more set of descriptive statistics and are not legitimate statistical tests. The segments were originally formed via cluster analysis based on precisely the same interest measures evaluated by the F ratios. Hence, the observed differences in interests across the fourteen segments are, by definition, not due to chance.[4] The needs data and the demographic and socioeconomic measures were not used as part of the input to the cluster analysis, however, and so the F ratios are independent of the method for creating the segmentation scheme.

Adult Male Concentration

The three segments in this category are named: Mechanics and Outdoor Life, Money and Nature's Products, and Family- and Community-Centered. Though all three are composed predominantly of adult males, their patterns of interests, needs, and demographic characteristics diverge considerably. For purposes of brevity, section titles in this chapter covering both demographic and socioeconomic results will be labeled "Demographics."

Mechanics and Outdoor Life

> Young adult, blue-collar males whose interests focus on noncompetitive activities emphasizing personal, physical accomplishment —e.g., auto repair, fishing, camping—interests that, by their very nature, do not require emphasis on interpersonal cooperation or support. High on needs for escape and unique/creative accomplishment.

Interests. People in this segment score higher than those in any other on their interest in Mechanical Activities, and lower than any other on their interest in Professional Sports. This latter finding is especially surprising in that it conflicts with the usual

(text continues on p. 58)

Table 3.2 Average Interest Factor Scores by Interest Segment[a]

	Entire Population[b]	Adult Male Concentration			Adult Female Concentration				Youth Concentration			Mixed				
		Mechanics and Outdoor Life	Money and Nature's Products	Family- and Community-Centered	Elderly Concerns	Arts and Cultural Activities	Home- and Community-Centered	Family-Integrated Activities	Competitive Sports and Science/Engineering	Athletic and Social Activities	Indoor Games and Social Activities	News and Information	Detached	Cosmopolitan	Self-Enrichment	Highly Diversified
Comprehensive News/Information	.09	.04	.35	.25	.43	.48	-.60	-.27	-.30	-.38	-.30	1.20	-1.22	.53	.42	
Athletic Activities—Participant	-.02	-.02	-.46	-.29	-.72	-.42	-.54	.07	.90	1.44	.55	-.71	-.42	.34	.76	
Household Activities and Management	.00	-.57	-.54	-.61	.18	-.23	.94	.92	-.65	-.25	.04	-.12	-.21	-.19	.61	
Classical Arts	.11	-.39	-.38	-.34	-.16	1.96	-.03	-.30	-.38	-.12	.17	-.60	-.16	1.30	.39	
Reaping Nature's Benefits	.00	.40	1.04	.82	-.20	-.38	-.26	-.12	-.41	-.20	-.38	-.35	-.30	-.11	.40	
Professional Sports	-.07	-.58	-.01	.30	-.49	.10	.13	-.40	.78	-.30	-.24	.67	-.19	-.41	.07	
Science and Engineering	-.05	-.14	-.29	.02	-.72	-.14	.22	-.59	.48	-.28	.25	.56	.09	-.09	-.37	
Popular Entertainment	.00	.32	.10	-.64	-.27	.08	.81	.13	.44	.61	.55	.29	-1.32	-.04	-.34	
Religion	-.15	-.63	-.26	.46	.52	.21	.02	-.46	-.36	.00	.46	-.29	-.26	-1.14	-.07	
Popular Social Issues	-.07	-.28	-.27	.00	-.78	-.06	-.29	.15	-.30	.56	-.14	.10	-.18	.41	.37	
Indoor Games	-.01	-.17	-.38	-.04	-.17	.23	-.49	.92	-.25	-.85	1.09	-.24	-.15	-.68	.31	
Community Activities	-.05	-.17	-.17	.31	-.44	-.27	.38	-.03	-.25	-.01	-.66	.00	.11	.09	-.18	
Investments	.02	.10	.86	.43	-1.25	.41	.29	.39	-.15	-.70	-.03	-.06	.05	-.30	-.05	
International Affairs	.04	.37	-.47	.05	-.42	.39	-.52	.06	.03	-.39	-.23	1.09	.29	.01	-.08	
Camping Out	.03	.51	-.27	.42	-.49	-.12	.01	.39	-.16	.06	.67	-.12	-.26	.95	.98	
Crime and Society	.01	-.01	-.93	.70	-.37	.20	.33	.43	-.02	.54	-.37	.06	-.26	-.20	.07	
Mechanical Activities	-.01	1.34	-.38	-.02	-.16	-.18	-.24	-.20	.63	-.58	-.89	-.44	-.17	-.38	-.78	
Contemporary Dancing	-.09	.27	-.17	-.52	-.42	-.18	-.24	-.25	-.99	.41	.74	.49	.25	-.37	.32	

a. The statistical significance of each of the eighteen interest factors contained in this table was evaluated based on univariate F ratios with 13 and 2462 degrees of freedom. All eighteen ratios were significant at the .005 level or better.
b. Average factor scores for entire sample based on unweighted data are zero by definition. However, reweighting of sample for projection to population results in nonzero values reported in this column.

Table 3.3 Average Need Factor Scores by Interest Segment[a]

	Entire Population[b]	Adult Male Concentration			Adult Female Concentration				Youth Concentration			Mixed			
		Mechanics and Outdoor Life	Money and Nature's Products	Family- and Community-Centered	Elderly Concerns	Arts and Cultural Activities	Home- and Community-Centered	Family-Integrated Activities	Competitive Sports and Science/Engineering	Athletic and Social Activities	Indoor Games and Social Activities	News and Information	Detached	Cosmopolitan Self-Enrichment	Highly Diversified
Socially Stimulating	-.01	-.06	.20	-.14	.26	.00	.17	-.04	-.10	.33	.22	.29	-.73	-.18	.05
Status Enhancement	-.10	.01	.11	.07	-.25	-.50	-.12	-.38	.21	.13	.34	-.11	.04	-.70	-.24
Unique/Creative Accomplishment	.01	.34	-.12	.13	-.57	-.03	-.25	.05	.33	.28	-.30	-.04	-.37	.45	-.22
Escape from Problems	-.01	.21	-.20	-.16	.10	-.29	-.13	.09	.00	.50	.10	-.19	-.09	.10	-.05
Family Ties	.02	-.11	.19	.48	-.11	-.04	.17	.38	-.06	-.52	-.10	.34	-.37	-.25	-.08
Understanding Others	.01	-.28	-.08	.04	.17	.42	.12	.10	-.68	.01	-.26	.08	-.09	.21	-.14
Greater Self-Acceptance	-.02	-.27	-.31	-.11	.22	.17	.11	.17	-.31	.11	.17	.18	-.43	.00	.12
Escape from Boredom	-.04	.22	-.11	-.21	.05	-.18	.02	-.11	.18	-.05	.14	.12	.00	-.53	-.07
Intellectual Stimulation and Growth	.02	.07	-.08	-.10	-.58	.33	-.16	.01	.21	.16	.11	.16	-.77	.70	.35

a. The statistical significance of each of the nine need factors in this table was evaluated based on univariate F ratios with 13 and 2462 degrees of freedom. All nine ratios were significant at the .005 level or better.
b. Average factor scores for entire sample based on unweighted data are zero by definition. However, reweighting of sample for projection to population results in nonzero values reported in this column.

Table 3.4 Demographic Characteristics by Interest Segment[a]

	Adult Male Concentration				Adult Female Concentration				Youth Concentration			Mixed				
	Entire Population	Mechanics and Outdoor Life	Money and Nature's Products	Family- and Community-Centered	Elderly Concerns	Arts and Cultural Activities	Home- and Community-Centered	Family-Integrated Activities	Competitive Sports and Science/Engineering	Athletic and Social Activities	Indoor Games and Social Activities	News and Information	Detached	Cosmopolitan	Self-Enrichment	Highly Diversified
Sex (percentage female)	52	4	23	17	71	69	84	87	5	83	91	43	47	59	51
Age (years)	40	29	53	47	61	44	44	35	22	19	22	47	46	36	34
Marital Status (in percentages)[b]																
Married	64	60	82	92	56	79	71	81	29	13	29	75	60	72	61
Widowed	8	—	10	2	35	10	10	3	1	—	3	10	15	3	6
Adults in Segment with Children																
Percentage	46	37	41	64	18	52	54	75	21	14	37	41	48	51	56
Children's age (mean years)	11	11	13	12	12	12	11	9	13	14	12	10	10	10	10
Race (in percentages)[c]																
Black	11	2	7	6	12	10	12	4	8	4	21	13	19	2	33
Spanish	4	—	1	—	3	4	3	4	3	6	12	2	5	8	7
White	83	93	92	93	83	86	84	91	86	87	66	85	74	86	56
Urban/Nonurban (in percentages)																
Central city	25	20	12	15	23	24	20	22	23	27	30	24	41	33	25
Suburb	41	38	35	39	30	49	48	49	51	43	36	40	36	45	33
Nonmetropolitan	34	42	53	46	47	27	32	29	26	30	34	36	23	22	42
Region (in percentages)																
Northeast	24	20	25	13	21	31	21	31	38	22	17	24	29	27	15
Central	24	25	18	23	25	19	25	34	19	31	26	29	26	21	19
South	35	40	41	44	40	26	33	27	27	29	34	40	38	23	49
West	17	15	16	21	14	24	21	9	16	17	23	7	8	29	18

a. The statistical significance of each of the demographic variables in this table was evaluated based on univariate F ratios with 13 and 2462 degrees of freedom. All but one, the Central region, were statistically significant at the .005 level or better.
b. Does not add to 100% due to omission of single, divorced, and separated.
c. Does not add to 100% due to omission of American Indian and other categories.

Table 3.5 Socioeconomic Characteristics by Interest Segment[a]

	Adult Male Concentration			Adult Female Concentration				Youth Concentration			Mixed			
	Mechanics and Outdoor Life	Money and Nature's Products	Family- and Community-Centered	Elderly Concerns	Arts and Cultural Activities	Home- and Community-Centered	Family-Integrated Activities	Competitive Sports and Science/Engineering	Athletic and Social Activities	Indoor Games and Social Activities	News and Information	Detached	Cosmopolitan Self-Enrichment	Highly Diversified
Income (in thousands of dollars)	14.9	14.6	14.2	6.9	16.6	14.3	16.1	16.6	16.8	11.7	14.3	10.6	18.8	12.7
Entire Population: 14.2														
Employment Status (in percentages)[b]														
Full time	64	41	63	17	46	25	36	31	25	17	48	39	45	52
Retired	4	39	23	40	17	18	5	3	–	–	23	22	3	3
Entire Population Full time: 40; Retired: 15														
Occupation (in percentages)														
Blue Collar	55	36	43	36	11	13	17	19	7	4	32	35	10	30
White Collar	3	13	17	21	31	29	31	5	23	18	32	16	25	20
Managerial/Professional	23	35	33	6	34	18	12	12	5	4	24	12	38	15
Homemaker	1	10	7	34	17	35	34	–	6	25	10	26	18	21
Student	17	4	4	3	8	4	5	63	60	49	2	10	8	14
Entire Population Blue Collar: 25; White Collar: 21; Managerial/Professional: 20; Homemaker: 19; Student: 15														
Education (in percentages)														
Grammar School	3	21	13	31	1	14	4	16	6	18	5	26	1	8
High School	63	44	48	60	30	62	68	66	74	67	58	50	34	62
College	29	35	39	9	69	24	28	18	20	15	37	24	65	30
Entire Population Grammar School: 12; High School: 55; College: 33														

a. The statistical significance of each of the socioeconomic variables in this table was evaluated based on univariate F ratios with 13 and 2462 degrees of freedom. All were significant at the .005 level or better.
b. Does not add up to 100% due to omission of part time, temporary, unemployed, armed forces, students, and homemakers.

association of males with athletic interests. These individuals are also well above average on their interest scores on both the Reaping Nature's Benefits and Camping Out factors.

Virtually all of the interests that load highly on the Reaping Nature's Benefits, Camping Out, and Mechanical Activities factors emphasize physical accomplishment in a noncompetitive mode. These include fishing, camping, and auto repair.

Though their interest in competition as modeled by professional sports is low, they show considerable interest in competitive forces in the international arena. Their interest score on International Affairs is well above average. All of the items that load highly on this factor relate to military, economic, and political competition, namely the arms race, the balance of trade, and conflict in the Middle East.

Needs. The members of this segment score well above average on their needs to Escape from Boredom and Escape from Problems, as well as on their need for Unique/Creative Accomplishment. At the opposite extreme, they score quite low on the needs for Understanding Others and for Greater Self-Acceptance. In general, their pattern of needs appears nonintrospective, and relatively individualistic (i.e., few needs are tied to relationships with other people).

Demographics. Fully 96% of this segment's members are male. They have an average age of 29 and, as such, are the youngest of the adult segments. They are more apt to be employed full time than the average for the population (64% versus 40%). In addition, they are much more likely than the other two adult male segments to have completed at least some high school, but less likely to have attended college.

Money and Nature's Products

> Older males with a higher proportion being rural and retired. Interests in passive activities that obtain some form of tangible return or product—fishing, hunting, investments. Low interest in active physical activities—camping out, participant sports—as well as culturally upscale or abstract—classical arts, international

affairs. Somewhat complacent, but need interpersonal contact and support, especially from their families.

Interests. Fourteen of the eighteen interest factor scores for people in this segment are below the average for the total population. Members of only two other segments, the Elderly Concerns and Detached segments, have as many interest scores below average.

Of the four interest factors ranked above average for this segment (Comprehensive News/Information, Reaping Nature's Benefits, Popular Entertainment, and Investments), two factors, Reaping Nature's Benefits and Investments, are especially high for this group. The one element these interests appear to have in common is that both involve a seeking of some tangible return for one's efforts.

In general, the interests of the members of this segment reflect relatively low levels of physical activity, perhaps because of their age. On those interest factors involving personal participation in relatively more strenuous activities, such as athletics, camping out, mechanics, and dancing, the members of this segment score well below the average for the total population.

They are also below average on their interest in Classical Arts, as well as some of the other more abstract intellectual content areas, such as Science and Engineering, International Affairs, and Crime and Society.

Needs. The need factors on which they are highest are those associated with social interaction and support, in both family and nonfamily contexts, namely Family Ties, Socially Stimulating, and Status Enhancement. On the other hand, they score well below average on Escape from Problems and Greater Self-Acceptance. These seem to be people whose needs are directed toward maintaining interpersonal contact and support, but who, at the same time, are relatively satisfied with themselves and their rather narrow span of interests and activities.

Demographics. The only segment with a higher average age than this one (53 years) is the Elderly Concerns segment, whose members average 61 years of age. Compared with members of the other adult segments, they are more likely to be married, and to live in suburban or rural areas. They are below average in education, but not in household income, despite the fact that a relatively large number are retired.

Their age, education, residence, and retirement characteristics are quite consistent with their interests and needs. They are disproportionately represented in both blue-collar and managerial and professional occupations.

Family- and Community-Centered

> Employed, blue-collar/white-collar adult males. Married, living in nonmetropolitan areas. Broad interests, including outdoor activities, investments, and home- and community-centered activities as well as religion. Very strong need for family ties.

Interests. The members of this segment have broad interests relative to the others in that they score above average on ten of the eighteen interest factors. This is matched by only one other group, namely the Highly Diversified segment.

The seven interest factors on which these people score the highest are: Camping Out, Crime and Society, Reaping Nature's Benefits, Religion, Community Activities, Investments, and Professional Sports. In addition, they have relatively high scores on Indoor Games, Popular Social Issues, and the Science and Engineering factors, compared to the members of the other two predominantly adult male segments.

Their high scores on Reaping Nature's Benefits and Investments constitute a level of interest similar to members of the Money and Nature's Products segment. However, their interests transcend those of the latter segment in that they are much broader.

One gets the impression that the Family- and Community-Centered people attempt to successfully integrate a diverse set of interests and activities with their home life. In contrast, the members of the Money and Nature's Products segment limit themselves to just a few interests that can be pursued on a more individualistic basis.

Needs. It is not surprising that the people in this segment score higher than any other on the need for Family Ties. Their lowest need scores are with respect to Escape from Problems and Escape from Boredom.

Demographics. As reflected in their family orientation, the adults in this segment are more apt to have children than are those in the population as a whole (64% versus 46%). Only one other segment has a higher proportion of adults with children and that is the Family-Integrated Activities segment (75%).

Family- and Community-Centered segment members are better educated than most and almost two-thirds are employed (63%). Fully 92% are married and 46% live in nonmetropolitan areas. Both of these figures are well above average for the population.

Adult Female Concentration

The four predominantly female segments are: Elderly Concerns, Arts and Cultural Activities, and Family Integrated Activities. The proportion of females in these segments ranges from 69% to 87%.

Elderly Concerns

Oldest segment, high percentage of retirees, widowed, few children. Very few interests include religion and news and information. Focus on maintaining sense of social integration and belonging in absence of direct interpersonal contact. Needs to overcome loneliness and lift spirits. Low need for intellectual stimulation.

Interests. The most striking characteristic of the members of the Elderly Concerns segment is that they score below the population average for fifteen of the eighteen interest factors. They are, however, higher than the members of any other segment on one interest factor, Religion. On the other, News and Information, they are the second highest segment. The only other interest factor on which they have an above average score is Household Activities and Management.

Needs. The members of this segment rank higher than any other on the need for Greater Self-Acceptance factor. This factor includes items involving overcoming loneliness and the need to lift one's spirits. They also score above average on the need to be Socially Stimulating. Also notable are the extremely low scores on the needs for Unique/Creative Accomplishment and for Intellectual Stimulation and Growth.

In other words, these people are struggling with the normal problems of aging. Unlike some who are able to take advantage of the freedoms offered by diminished work and family responsibilities by broadening their horizons, the members of this segment appear to be somewhat passive in accepting and adjusting to a more circumscribed life style as they turn their world inward upon themselves.

Demographics. People in this segment average 61 years of age. They include, by far, the highest percentage of retirees, 40%, versus 15% for the population. Fully 35% of them are widowed. Only 18% are adults with children remaining at home. Slightly fewer than one-third (31%) have no more than a grammar school education.

Arts and Cultural Activities

Highly educated, adult women in households with manager or professional as head. Broad range of intellectual and cultural interests—especially classical arts. Low interest in household activities and management. High needs for intellectual stimulation and growth and for understanding others with low needs for status enhancement and escape.

Interests. The members of this segment rank considerably higher than any other on the Classical Arts factor. They report a broad range of interests, scoring above average on thirteen of the eighteen interest factors, including interest scores that place them among the top three segments for Investments, International Affairs, and Comprehensive News/Information. They are well below average in their interest scores in Participant Athletic Activities, Community Activities, Reaping Nature's Benefits, and report by far the lowest interest level in Household Activities and Management of the four adult female segments.

Needs. These people are below average on their need scores for Status Enhancement, Escape from Problems, and Escape from Boredom. In contrast, they rank higher than any other segment on the need for Understanding Others and are among the three highest segments in their need for Intellectual Stimulation and Growth.

Demographics. The pattern of their demographic characteristics is quite consistent with that of their interests and needs. Some 69% of them have at least some college education, compared to 33% for the population, and 65% of those employed are either managers, professionals, or white-collar workers. They are overrepresented in the Northeast and West and underrepresented in the central and southern regions of the country. They also tend more than the rest of the population to live in suburban areas and are less often found in nonmetropolitan areas. A large proportion, 79%, are married, compared to 64% for the population.

Home- and Community-Centered

> Adult females with a relatively high percentage of married homemakers. Home and local community interests. Highest needs for family ties and understanding others. Lowest needs for intellectual stimulation and for unique/creative accomplishment.

Interests. People in this segment rank higher than those in any other on three interest factors: Household Activities, Popular Entertainment, and Community Activities. The entertainment activities associated with the Popular Entertainment factor, such as visiting friends, radio, travel/sightseeing, popular music, dining out, and movies, are in marked contrast to those entertainment activities associated with the Classical Arts factor, which include opera, classical music, ballet, and live theater.

The Home- and Community-Centered segment members are well below average on their interest scores for International Affairs, Comprehensive News/Information, Participant Athletic Activities, Indoor Games, and Popular Social Issues. With respect to all of these interests, they score among the lowest three segments.

Needs. None of their need scores are extremely high or low in relation to the other segments. Nonetheless, their highest and lowest need scores are consistent with their interests. The highest two scores for people in this segment are for the need for Family Ties and the need to be Socially Stimulating. Their lowest scores are for Unique/Creative Accomplishment and for Intellectual Stimulation and Growth.

This segment appears to be composed primarily of women who, even in today's changing world, are filling the role of homemaker in a rather narrow, traditional manner.

Demographics. A below average proportion of the members of this segment are employed full time (25% versus 40% for the population). Fully 35% of the members of this segment are homemakers compared to the 19% average across all groups. With these exceptions, none of their demographics tend to clearly distinguish them from the other predominantly adult female segments.

Family-Integrated Activities

> High percentage of adult women with young children. Strong interest in home and in family interactive activities—household activities and management and indoor games. High need for family ties. Child presence influences adult interest patterns.

Interests. As in the case of the Home- and Community-Centered segment, the members of this segment are also well above average on their interest score for Household Activities. However, the interests of this segment's members are somewhat broader. They are also well above average with respect to their score on Investments, Camping Out, and Crime and Society. In addition, they exhibit a much stronger interest in Comprehensive News/Information, but a lesser interest in Professional Sports and in Science and Engineering. The Family-Integrated Activities segment members are sharply differentiated from the Home- and Community-Centered segment by the former's strong interest in Indoor Games.

Needs. People in this segment are well above average on their need score for Family Ties and well below average on their need score for Status Enhancement.

Demographics. Fully 87% of the people in this segment are women. The only other predominantly adult female segment that has near this proportion of female members is the Home- and Community-Centered segment, which consists of 84% women. Of the adults in this segment, 75% have children, by far the largest percentage of any of the segments. Their average age is 35, the youngest of the adult female groups. Their children are also on average younger (9 years of age) than those of any other segment. Members of this segment are disproportionately white and tend to reside in suburban areas (49% versus 41% for the entire sample) and in the central United States (34% versus 24%).

Youth Concentration

The three segments in this category are labeled as follows: Competitive Sports and Science/Engineering, Athletic and Social Activities, and Indoor Games and Social Activities. Their members range in age from an average of 19 to 22 years. The first segment is 95% male, while the latter two are 83% and 91% female, respectively.

Competitive Sports and Science/Engineering

Teenage male students with interests in male-associated mechanical activities and competitive athletics. Avoidance of female-oriented subjects and interests. High on needs for unique/creative accomplishment, intellectual stimulation and growth, status enhancement, and escape from boredom. Low needs for understanding others and for greater self-acceptance.

Interests. There are four interest factors on which this segment's members are among the top two in the study. They are Professional Sports, Participant Athletic Activities, Science and Engineering, and Mechanical Activities. At the opposite extreme there are five interest factors on which they rank among the bottom two, namely: Household Activities, Classical Arts,

Reaping Nature's Benefits, Popular Social Issues, and Contemporary Dancing.

The people in this segment are attracted by interests that involve traditional male-related, competitive interactions and those associated with competence in mechanical or technological areas. They have little interest in activities or subjects that tend to be associated with female roles, such as Household Activities and Contemporary Dancing. They also exhibit low levels of interest in current events, as evidenced by their low scores on Comprehensive News/Information and Popular Social Issues. They are also less attracted to more abstract subject matters such as the Classical Arts.

Needs. The need scores for people in this segment are above average for Status Enhancement, Unique/Creative Accomplishment, Intellectual Stimulation and Growth, and Escape from Boredom. They are among the bottom two segments on the Understanding Others and Greater Self-Acceptance factors.

The segment is composed largely of adolescent boys whose interest patterns appear to be rather narrow and recreational. They have not yet begun to expand their horizons into more abstract, intellectually oriented content areas. Their pattern of needs reflects the adolescent struggle for identity as they prepare to move into the world of adults.

Demographics. Consistent with their interests and needs, the members of this segment are 95% male with an average age of 22 and the largest proportion of students (63%). They disproportionately reside in the suburbs and in the Northeast.

Athletic and Social Activities

> Teenage females from high-income families. The youngest of all the segments. Interests in active, away-from-home, face-to-face activities. High need to escape from problems and to be socially stimulating. Low need for family ties.

Interests. There are five interest factors on which this segment is well above average, namely: Participant Athletic Activities,

Popular Entertainment, Contemporary Dancing, Popular Social Issues, and Crime and Society. It is our hypothesis that the common denominator of this set of interests and subjects is that they involve active, face-to-face interaction in nonhome settings, especially the first three interests mentioned. The latter two represent subject matter that may provide the basis for interpersonal conversation among their peer group.

Interests such as Indoor Games, which tend to be home-based vehicles for face-to-face interaction, score well below average. People in this segment rank lowest on this interest score. Interests that extend beyond their immediate sphere of influence, such as Comprehensive News/Information, Investments and International Affairs receive interest scores well below average.

Needs. The importance of peer group relationships to the members of this segment is further signaled by their scores on the need to be Socially Stimulating (on which they have higher scores than any other segment), and on the Status Enhancement factor, on which their score is above average, ranking third.

The away-from-home orientation of their interests is consistent with the fact that their need score for Family Ties is lower than that for any other segment, and their score on the need to Escape from Problems is higher than that for any other segment. The highest loading item on this latter factor is: "To get away from the pressures and responsibilities of my home life."

This segment contains a large number of active adolescent girls who appear to be breaking away from the pressures of home and family and are channeling their energy into physical and social activities with their peers.

Demographics. This segment is 83% female. Besides their average age (19 years) and related characteristics, the only other demographic or socioeconomic characteristic that sharply differentiates them is the average family income of the households to which they belong, $16,800, which compares to $14,200 for the entire sample and ranks the members of this segment among the top two segments.

Indoor Games and Social Activities

Young, low-income, females. Interests in activities, especially indoor games. Low interest in most subject matter areas. Nonintellectual. High needs for status enhancement and the need to be socially stimulating.

Interests. The members of this segment share with those in the Athletic and Social Activities segment strong interests in Popular Entertainment and Contemporary Dancing and low interest in Mechanical Activities.

The single interest factor that most sharply differentiates people in this segment from those in the Athletic and Social Activities segment is Indoor Games. The members of this segment have the highest score of any segment on this factor, while those in the Athletic and Social Activities segment have the lowest. The next most important interest factor differentiating them is Participant Athletic Activities on which they score much lower than the other segment. This segment has no high interest scores on any of the factors that emphasize intellectual content, although they exhibit by far the highest interest in Religion among the youth groups.

Needs. Like the Athletic and Social Activities segment, this group scores high on the need for Status Enhancement and the need to be Socially Stimulating. The Indoor Games and Social Activities segment, however, is distinguished by much lower needs for Unique/Creative Accomplishment and for Understanding Others. The latter group has a below-average need for Family Ties, but not nearly to the extreme degree as the former group.

Demographics. This segment's members are young (average age 22) and primarily female (91%). Their lack of a strong away-from-home orientation is, at least in part, due to the fact that 37% of the members of this segment are adults in families with children, compared to 14% for the previous segment.

In addition, the average family income of members of this segment is $11,700 versus $16,800 for those in the Athletic and

Social Activities segment. The people in this segment are also disproportionately Black (21% versus 11% for the population) or Hispanic (12% versus 4% for the population).

Compared to the previous segment, whose members are primarily older adolescents, the women in the Indoor Games and Social Activities segment tend to be young adults. Their interests and needs reflect the changing patterns associated with the age differential. There is less of a need to focus on seeking independence from parental authority and more of a need, having gained some of the independence, to establish a new household and an identity as an adult household head.

Mixed

These segments include: News and Information, Detached, Cosmopolitan Self-Enrichment, and Highly Diversified.

The four segments discussed in this section are composed predominantly of adults whose average age ranges from 34 to 47. The composition of males and females in these segments is much more balanced than that of the ten segments previously discussed. For example, the proportion of females in these four segments ranges from 43% to 59%, whereas in the four predominantly adult female segments it ranged from 71% to 87%. The corresponding proportions of males are 41% to 57% among the four segments to be discussed and 77% to 96% among the three predominantly adult male segments previously presented.

News and Information

> Passive interests related to keeping informed on a broad range of subjects and activities. Needs are focused on being socially stimulating and maintaining family ties.

Interests. As the label suggests, the people in this segment are interested in being informed on a wide range of subjects. They rank among the top two segments on the following five interest factors: Comprehensive News/Information, International Affairs, Professional Sports, Science and Engineering, and Contemporary Dancing. The interests of this segment are quite broad in relation to subject matter. They do not, however, encompass the more abstract, cultural interests, as evidenced by their score on the Classical Arts factor, placing them as the least interested of all the segments.

In addition, they appear to be more oriented toward observing and knowing, as opposed to directly participating. The only segment whose members rank lower on the Participant Athletic Activities factor is the Elderly Concerns segment. They also rank well below average on the Reaping Nature's Benefits factor.

Needs. This segment's members exhibit relatively high scores on the need for Family Ties and the need to be Socially Stimulating. Their scores are not very low on any of the other seven needs measured. Their lives appear to center around making themselves both collectors and distributors of a wide variety of information.

Demographics. For the most part neither the demographic nor the socioeconomic characteristics of this segment tend to set it apart from the others. They have an average age of 47, and 57% percent of them are male. Their employment status is quite diverse, reflecting the relatively heterogeneous age and sex composition of the segment. They do have the highest proportion of white-collar workers (32%) among all fourteen segments.

Detached

Low socioeconomic profile. Extremely few interests and activities and few psychological needs satisfied by them. Low scores on needs related to both intellectual stimulation and interpersonal contact and support.

Interests. People in this segment score well below average on thirteen of the eighteen interest factors. They do not rank first or second on any factor. They rank third on only one, Community Activities, and fourth on another, International Affairs. Members in only two other segments have a span of interests as narrow, namely the Elderly Concerns and the Money and Nature's Products segments, with fifteen and fourteen interest factor means below average, respectively. In both cases, these segment members have at least one interest factor on which they score higher than all other segments. The Detached segment's members not only have a narrow range of interests, they also tend not to score as high on the few interests that they do rate above average. In addition, they are the lowest-scoring segment on their interest in both Comprehensive News/Information and Popular Entertainment. That is, they show little evidence of any desire to stay in touch with the day-to-day events in the world that surrounds them.

Needs. The profile of need factor scores for people in this segment reveals a picture consistent with their interests. On none of the nine need factors do the members of this segment rank higher than sixth out of the fourteen segments. The two needs on which they have the highest relative position are Status Enhancement and Escape from Boredom. For five of the nine need factors, they have the lowest, or second lowest, need scores of all the segments. The five are Socially Stimulating, Unique/Creative Accomplishment, Family Ties, Greater Self-Acceptance, and Intellectual Stimulation and Growth. The members of the Detached segment have few, if any, strongly felt needs that we have been able to identify.

Demographics. This segment has the highest proportion of central-city residents (41% versus 25% for the population). They are about evenly divided between men and women and above average in age (46 years versus 40). They rank third in their proportion of Black members (19% versus 11%) with a higher proportion being widowed (15% versus 8%).

They have the second lowest average income ($10,600 versus $14,200) and the second highest proportion of members who did not complete high school (26% versus 12%).

It appears that this segment comprises those members of society who are quite low on the socioeconomic ladder and whose interests, activities, and needs are more fundamental than those investigated in this study. The types of leisure interests and psychological needs investigated in this project are probably of little relevance to a group that is concerned with providing the more basic needs of food, clothing, and shelter for themselves and their families.

Cosmopolitan Self-Enrichment

Extremely high socioeconomic profile. Diverse pattern of intellectual and cultural interests. Physically active. High needs for intellectual stimulation, unique/creative accomplishment, and understanding others. Low needs for status enhancement and for escape from boredom.

Interests. Members of this segment and the Arts and Cultural Activities segment have in common exceptionally high interest scores on the Classical Arts and Comprehensive News/Information factors. They share an interest in abstraction and culturally upscaled subject matter.

The differences between people in the two segments are as informative as their similarities. Based on a comparison of their relative positions among the fourteen segments:

(1) The members of the Arts and Cultural Activities segment rank notably higher than those in the Cosmopolitan Self-Enrichment segment with respect to their scores on the following interest factors: Professional Sports, Religion, Indoor Games, Investments, International Affairs, and Crime and Society.
(2) The members of the Cosmopolitan Self-Enrichment segment rank notably higher than those in the Arts and Cultural Activities segment on: Participant Athletic Activities, Popular Social Issues, Community Activities, and Camping Out.

The Cosmopolitan Self-Enrichment segment's members have interests that appear to involve more active participation than do those in the Arts and Cultural Activities segment. The mix of their interests spans not only those related to intellectualizing and abstraction but also those that are people- and community-related. The members of the Cosmopolitan Self-Enrichment segment, in general, appear to have a set of interests that are more highly crystalized and require a greater sense of involvement and participation.

Needs. Members of both segments are well above average, among the top two, on their need scores related to Intellectual

Stimulation and Growth and Understanding Others. They are both well below average on their need scores for Status Enhancement and Escape from Boredom. They diverge, however, in that the Cosmopolitan Self-Enrichment segment members are (1) the highest ranking on the Unique/Creative Accomplishment factor, versus a slightly below average score for the members of the Arts and Cultural Activities segment, and (2) above average, ranking third, on the Escape from Problems factor, while the Arts and Cultural Activities segment ranks last.

In contrast, people in the Arts and Cultural Activities segment score higher than those in the Cosmopolitan Self-Enrichment segment on such people-related needs as Socially Stimulating, Family Ties, and Greater Self-Acceptance.

Demographics. Common to the members of both segments is an exceptionally high educational level, namely 65% with some college or more for the Cosmopolitan Self-Enrichment segment and 69% for the Arts and Cultural Activities segment. Both segments also contain a disproportionate number who are managerial or professional. They both comprise people with household incomes well above average, $18,760 for members in the Cosmopolitan Self-Enrichment segment and $16,570 for those in the Arts and Cultural Activities segment. The former segment, however, averages 36 years of age, while the latter averages 44 years.

This segment comprises a group of well-educated, relatively affluent men and women whose interests and activities include a broad spectrum of intellectual and cultural areas. They are active. They participate in their family and community, although not in religion.

Highly Diversified

> Southern, Black, adults with children. Broad range of interests, especially those permitting personal participation with family and/or other informal small group settings. High need for intellectual stimulation and growth.

Interests. Of the eighteen interest factors, the people in this segment are above average on eleven. No segment has a larger number of above average interest factor scores than this one. The two closest segments are the Family and Community-Centered segment, with the same number, and the Family-Integrated Activities segment with ten.

The members of this segment rank first on only one activity, namely Camping Out, but they rank well above average on the following six: Participant Athletic Activities, Household Activities, Reaping Nature's Benefits, Classical Arts, Popular Social Issues, and Indoor Games. With the exception of Popular Social Issues, all of the interests on which they score high involve some form of active personal participation.

There are three interest factors on which the people in this segment score well below average, namely: Science and Engineering, Mechanical Activities, and Popular Entertainment. The first two of these factors involve interests in technical subjects. The third tends to be highly oriented toward today's youth with low intellectual or cultural content. The interests of this group tend to be either more intellectually upscale or to involve some form of active participation.

Needs. This segment reports a high need for Intellectual Stimulation and Growth and relatively low needs on the remaining factors.

Demographics. The Highly Diversified segment is most notable demographically for its large proportion of Blacks (33% compared to 11% in the population), by far the largest of all the segments. Geographically it includes the largest representation from the South and the second smallest from the Northeast. Full-time employment is among the highest of all the segments, although household income averages only $12,700 compared to $14,200 for the population.

This segment appears to comprise a group of people, who, without the benefit of the educational and social advantages of others in the population, are seeking to expand the intellectual and cultural horizons of themselves and their families.

Conclusions

The fourteen interest segments that have been introduced in this chapter are, in subsequent chapters, used as the focal point for the analysis of public television's audience. We believe that this particular interest-based segmentation scheme provides a useful complement to the more traditional demographic segmentation based on age and sex that is often used for the analysis of public, as well as commercial, television. Our previous book (Frank and Greenberg, 1980) documents in detail the relationship between segment membership and usage of television in general, and other media such as books, magazines, movies, newspapers and radio. Our interest segmentation is a tool that can be used to help understand the audience for PTV and thereby help, in a modest way, those concerned with the future development of PTV.

Figure 3.1 provides a summary of the fourteen interest segments. They have been put together at the end of the chapter so they can be used as a convenient reference when reading the chapters that follow.

Notes

1. Except for the concluding section, this chapter is virtually identical to Chapter 4 in Frank and Greenberg (1980). Knowledge of the interest segments is crucial for interpreting the findings in the chapters that follow; hence, the necessity for including this material. Without it the reader would be forced to read our previous book in order to interpret the findings in this one.

2. For a more detailed comparison of our interest segmentation with a demographic one see Frank and Greenberg (1980, Chapter 10).

3. For a more detailed discussion of the distinction between people type analysis and variables analysis see Frank and Massy (1975).

4. Fourteen-way multiple discriminant analyses were also performed as part of the process of evaluating the extent of the overall agreement between segment differences in interests, needs, demographic and socioeconomic characteristics. Three discriminant analyses were performed, each using a different set of variables as follows: (1) interest factor scores, (2) need factor scores, and (3) demographic and socioeconomic characteris-

(Notes continue p. 78)

Figure 3.1
Interest Segmentation Sketches

Adult Male Concentration

(1) *Mechanics and Outdoor Life.* Young, adult, blue-collar males whose interests focus on noncompetitive activities emphasizing personal physical accomplishment—e.g., auto repair, fishing, camping, interests that, by their very nature, do not require emphasis on interpersonal cooperation or support. High on needs for escape and unique/creative accomplishment.

(2) *Money and Nature's Products.* Older males with a higher proportion being rural and retired. Interests in passive activities that obtain some form of tangible return or product—fishing, hunting, investments. Low interest in active physical activities—camping out, participant sports—as well as culturally upscale or abstract—classical arts, international affairs. Somewhat complacent, but need interpersonal contact and support, especially from their families.

(3) *Family- and Community-Centered.* Employed blue-collar/white-collar adult males. Married, living in nonmetropolitan areas. Broad interests, including outdoor activities, investments, and home- and community-centered activities as well as religion. Very strong need for family ties.

Adult Female Concentration

(1) *Elderly Concerns.* Oldest segment, high percentage of retirees, widowed, few children. Very few interests include religion and news and information. Focus on maintaining sense of social integration and belonging in absence of direct interpersonal contact. Needs to overcome loneliness and lift spirits. Low need for intellectual stimulation.

(2) *Arts and Cultural Activities.* Highly educated, adult women in households with manager or professional as head. Broad range of intellectual and cultural interests—especially classical arts. Low interest in household activities and management. High needs for intellectual stimulation and growth and for understanding others with low needs for status enhancement and escape.

(3) *Home- and Community-Centered.* Adult females with a relatively high percentage of married homemakers. Home and local community interests. Highest needs for family ties and understanding others. Lowest needs for intellectual stimulation and for unique/creative accomplishment.

(4) *Family-Integrated Activities.* High percentage of adult women with young children. Strong interest in home and in family interactive activities—household activities and management and indoor games. High need for family ties. Child presence influences adult interest patterns.

AUDIENCE INTEREST SEGMENTATION

Figure 3.1 Continued

Young Concentration

(1) *Competitive Sports and Science/Engineering.* Teenage male students with interests in male-associated mechanical activities and competitive athletics. Avoidance of female-oriented subjects and interests. High on needs for unique/creative accomplishment, intellectual stimulation and growth, status enhancement, and escape from boredom. Low needs for understanding others and for greater self-acceptance.

(2) *Athletic and Social Activities.* Teenage females from high-income families. The youngest of all the segments. Interests in active, away-from-home, face-to-face activities. High need to escape from problems and to be socially stimulating. Low need for family ties.

(3) *Indoor Games and Social Activities.* Young, low-income females. Interests in activities, especially indoor games. Low interests in most subject matter areas. Nonintellectual. High needs for status enhancement and the need to be socially stimulating.

Mixed

(1) *News and Information.* Passive interests related to keeping informed on a broad range of subjects and activities. Needs are focused on being socially stimulating and maintaining family ties.

(2) *Detached.* Low socioeconomic profile. Extremely few interests and activities and few psychological needs satisfied by them. Low scores on needs related to both intellectual stimulation and interpersonal contact and support.

(3) *Cosmopolitan Self-Enrichment.* Extremely high socioeconomic profile. Diverse pattern of intellectual and cultural interests. Physically active. High needs for intellectual stimulation, unique/creative accomplishment, and understanding others. Low needs for status enhancement and for escape from boredom.

(4) *Highly Diversified.* Southern, Black, adults with children. Broad range of interests, especially those permitting personal participation with family and/or other informal small group settings. High need for intellectual stimulation and growth.

tics. The interest-based discriminant analysis was included simply to provide another set of descriptive statistics and not to evaluate any null hypotheses for the same reasons cited in the text. Based on interest factor scores, 94% of the respondents were correctly classified based on the discriminant equations, compared to the 7% one could expect from a random assignment to the fourteen segments, or the 10% accuracy that would occur if all were assigned to the modal segment. One, of course, would expect this percentage to be extremely high, as the same interest factors used in the discriminant analysis also served as input to the cluster analysis that formed the fourteen segments in the first place. The need and demographic/socioeconomic discriminant runs resulted in 23% and 33% being correctly classified, which is substantially greater than the chance expectations of 7% or 10%. The overall pattern of between-segment differences for both of these discriminant analyses were tested using the Wilk's Lambda statistic. Based on the appropriate F-ratio approximation, the results for both discriminant analyses were significant beyond the .005 level.

4

Public Television Viewing Behavior

This chapter reports our findings regarding PTV viewing behavior for each of the fourteen interest segments. First to be discussed is an analysis of PTV viewing behavior for the entire population, followed by individual discussions of each of the interest segments.

PTV Viewing Behavior by the Entire Population

Figure 4.1 reports the percentage of the population who indicate varying degrees of PTV watching. Less than half (47.2%) report usually watching PTV once a year or more, while 46.5% report never watching it.

In the second Carnegie Commission report, two audience coverage objectives are recommended (1979: 276). They propose that PTV should seek to serve all Americans. In the short run, they recommend that a goal be set of reaching each person in the population at least once a month. They also foresee a potential goal of at least once a week rather than once a month. By either criterion, PTV has a long way to go. A total of 35.5% report usually watching once a month or more, while only 25.7% report usually watching one or more times per week. Against these goals, PTV is still faced with the potential challenge of reaching between two-thirds and three-quarters of the U.S. population aged 13 and over.

Figure 4.1
Extent of Public Television Viewing for the Entire Population

Every day	7.0%	} 25.7% Frequent	}
One to six times per week	18.7		} 47.2% PTV Viewers
Once or twice a month	9.8	}	
A few times a year	9.4	} 21.5% Occasional	
Once a year	2.3	}	
Don't know	4.9	}	
Never	46.5	} 52.8% Never/Undetermined	
No answer	1.4	}	
Entire population	100.0%		

The ability to reach such broad audiences is currently constrained by the relatively light viewership of any given PTV program. Respondents were asked how often in the last four weeks they watched each of 150 television programs, a list that included 22 programs aired on PTV. The proportion of the entire population reporting having watched any one particular program at least once averaged 4.6% and ranged from a low of .8% for *In Pursuit of Liberty* to a high of 7.0% for *Evening at Symphony*.[1] With few exceptions PTV's program audiences are smaller than the minimum audience size necessary to keep a program series on commercial television.

While these PTV viewing figures are relatively low taken over the entire population, there is substantial variation in the extent to which members of the fourteen interest segments are reached by PTV.

PTV Viewing Behavior by Interest Segment

Background Details

Tables 4.1, 4.2, and 4.3 report our findings on the PTV viewing behavior of members of each of the fourteen interest segments.

Table 4.1 Public Television Exposure Index by Interest Segment (in percentages)

	Public Television Exposure Index[a]
Arts and Cultural Activities (AF)[b]	200
Cosmopolitan Self-Enrichment (M)	192
News and Information (M)	120
Highly Diversified (M)	115
Family-Integrated Activities (AF)	105
Athletic and Social Activities (Y)	97
Home- and Community-Centered (AF)	88
Mechanics and Outdoor Life (AM)	83
Indoor Games and Social Activities (Y)	79
Competitive Sports and Science/Engineering (Y)	66
Detached (M)	61
Money and Nature's Products (AM)	57
Elderly Concerns (AF)	50
Family- and Community-Centered (AM)	48
Entire population (percentage watching once a week or more)	(25.7)

a. All percentages are indexed against a figure of 100% for the average of the entire population.
b. Letters associated with each segment indicate which concentration it is in, namely: AF = Adult Female, AM = Adult Male, Y = Youth, and M = Mixed.

This section describes what is measured by each of these tables. As in the preceding chapter we have deemphasized the citing of specific tables and values within them. The tables have been retained in the body of the text so that the reader who wishes to check our reasoning against the data on which it is based may do so.

Tables 4.1 and 4.2 summarize data concerning the extent to which the fourteen interest segments report usually watching PTV. Table 4.1 reports an index that reflects the proportion of a segment's members who watch PTV once a week or more compared to that for the entire population (25.7%). For example, the rate of 200% reported for people in the Arts and Cultural Activities segment indicates that the proportion of people in this segment who reported watching PTV once a week or more (51.4%) is twice that of the population as a whole. At the other

Table 4.2 Frequency of PTV Viewing by Interest Segment

	Entire Population	Frequent	Occasional	Never/ Undetermined
(a) Across-Segment Percentage Distribution of PTV Viewing				
Arts and Cultural Activities (AF)[a]	8.7	17.7	8.2	4.6
Cosmopolitan Self-Enrichment (M)	7.9	15.3	12.5	2.4
News and Information (M)	5.4	6.2	5.1	5.1
Highly Diversified (M)	8.0	9.1	7.2	7.8
Family-Integrated Activities (AF)	9.9	10.4	12.2	8.8
Athletic and Social Activities (Y)	4.2	4.1	4.4	4.2
Home- and Community-Centered (AF)	7.8	6.9	8.3	8.0
Mechanics and Outdoor Life (AM)	8.3	6.9	9.0	8.6
Indoor Games and Social Activities (Y)	4.0	3.2	2.9	4.9
Competitive Sports and Science/Engineering (Y)	6.5	4.4	7.9	7.0
Detached (M)	8.6	5.2	5.4	11.6
Money and Nature's Products (AM)	6.2	3.5	5.8	7.7
Elderly Concerns (AF)	8.0	3.9	4.4	11.5
Family- and Community-Centered (AM)	6.4	3.1	6.7	7.9
Entire population	100	100	100	100
(b) Within-Segment Percentage Distribution of PTV Viewing				
Arts and Cultural Activities (AF)		52.2	20.1	27.7
Cosmopolitan Self-Enrichment (M)		50.1	34.0	15.9
News and Information (M)		30.2	20.2	49.6
Highly Diversified (M)		29.4	19.2	51.4
Family-Integrated Activities (AF)		26.9	26.5	46.6
Athletic and Social Activities (Y)		25.0	22.5	52.6
Home- and Community-Centered (AF)		22.8	23.1	54.1
Mechanics and Outdoor Life (AM)		21.5	23.4	55.1
Indoor Games and Social Activities (Y)		20.4	15.3	64.3
Competitive Sports and Science/Engineering (Y)		17.3	26.1	56.6
Detached (M)		15.6	13.5	70.9
Money and Nature's Products (AM)		14.6	20.1	65.3
Elderly Concerns (AF)		12.5	11.7	75.9
Family- and Community-Centered (AM)		12.3	22.6	65.0
Entire population		(25.7)	(21.5)	(52.8)

a. Letters associated with each segment indicate which concentration it is in, namely: AF = Adult Female, AM = Adult Male, Y = Youth, and M = Mixed.

extreme, the index of 48% for members of the Family- and Community-Centered segment indicates that the proportion of frequent PTV viewers in that segment (12.3%) is less than half that found in the population as a whole.

Table 4.2a presents the distribution of people across all fourteen segments for each level of PTV viewing (frequent, occasional and never/undetermined). For example, the Arts and Cultural

Activities segment members account for 8.7% of the entire sample, 17.7% of frequent PTV viewers, 8.2% of occasional viewers, and only 4.6% of those in the never/undetermined category. The frequent, occasional, and never/undetermined category definitions are the same as those reported in Table 4.1.

Table 4.2b reports the proportional compositions of each interest segment in terms of frequent, occasional, and never/undetermined PTV viewing. For example, only 15.9% of those in the Cosmopolitan Self-Enrichment segment are in the never/undetermined category, while 70.9% of those in the Detached segment are in it.

The interest segments in both of these tables are listed in order of the percentage of people in them who watch PTV once a week or more. This ranking facilitates comparisons among segments who are above and below average in their PTV usage.

Table 4.3 reports the viewing data for each of the 22 PTV programs included in the survey by interest segment. It summarizes the extent to which the people in each segment serve as a source of each program's audience. For example, those in the Arts and Cultural Activities segment accounted for 47.8% of the audience viewing *Dickens of London*. In contrast, those in the Athletic and Social Activities segment accounted for only 1.6% of its audience.

Above-Average Viewing Segments

In five of the fourteen segments the proportion of people watching PTV once a week or more is above the population average of 25.7%. These segments, together with their PTV Exposure Indices (Table 4.1), are as follows:

- Arts and Cultural Activities (Adult Female Concentration)—200% of average or 100% above average
- Cosmopolitan Self-Enrichment (Adult Male Concentration)—192%
- News and Information (Adult Male Concentration)—120%
- Highly Diversified (Adult Male Concentration)—115%
- Family-Integrated Activities (Adult Female Concentration)—105%

(text continues p. 86)

Table 4.3 Segment Shares of PTV Program Viewing by Interest Segment (in percentages)

Type of Program	Arts and Cultural Activities (AF)[a]	Cosmopolitan Self-Enrichment (M)	News and Information (M)	Highly Diversified (M)	Family-Integrated Activities (AF)	Athletic and Social Activities (Y)	Home- and Community-Centered (AF)	Mechanics and Outdoor Life (AM)	Indoor Games and Social Activities (Y)	Competitive Sports and Science/Engineering (Y)	Detached (M)	Money and Nature's Products (AM)	Elderly Concerns (AF)	Family- and Community-Centered (AM)	Entire Population
Documentary															
Age of Uncertainty	21.1*	11.3	15.1*	12.1	8.0	–	–	2.4	1.1	16.8*	1.9	–	10.2	–	100
Nova	22.0*	16.6*	2.9	8.3	15.3*	.4	7.6	2.8	.3	5.7	5.4	3.5	7.1	2.0	100
Theatrical Performances															
Dickens of London	47.8*	27.1*	2.3	3.0	–	1.6	6.5*	2.3	1.6	.6	3.1	3.0	1.0	.1	100
Visions	34.6*	16.4*	13.3*	10.6	–	1.2	8.8	.1	3.5	.7	9.5	–	.9	.3	100
Upstairs, Downstairs	41.4*	15.2*	1.5	6.3	.5	.2	1.7	2.4	1.6	.2	4.1	5.0	10.8*	9.0	100
In Pursuit of Liberty	15.8*	2.2	32.8*	14.2*	12.9	–	10.0	.2	1.6	–	3.3	–	3.3	3.8	100
Great Performances	41.4*	15.9*	7.9	11.5*	.7	.3	1.4	.7	2.1	–	2.1	7.0	7.3	1.8	100
The Best of Families	19.8*	8.6	16.8*	13.6*	11.0	.1	.4	3.2	5.2	1.0	8.5	3.5	8.4	–	100
Masterpiece Theatre	32.5*	11.9*	3.7	4.6	11.4*	.3	5.4	3.5	3.0	3.3	2.2	5.1	7.5	5.8	100
Once Upon a Classic	17.5*	31.4*	2.9	10.5	18.9*	3.0	.2	2.8	1.0	1.4	4.6	3.2	.3	2.1	100

Musical Performances															
Evening at Symphony	40.9*	15.9*	2.2	3.1	4.2	—	6.0	5.7	1.5	.3	4.0	7.5*	7.4	1.1	100
Opera	51.3*	9.2*	.8	8.5*	7.9	.1	7.4	.2	1.9	—	3.6	4.1	5.0	—	100
Evening at Pops	33.5*	19.4*	6.4*	6.0	4.6	—	4.9	4.6	1.8	1.0	6.3	3.8	4.9	2.9	100
News/Documentary															
Washington Week in Review	22.6*	6.0	14.3*	8.3	1.8	—	10.3	3.0	.1	.5	4.5	16.6*	5.7	6.2	100
Wall Street Week	32.6*	5.3	19.6*	5.2	—	—	7.7	7.6	.7	—	4.0	9.6*	—	7.2	100
Black Perspective on the News	9.3	5.3	13.9*	27.0*	2.7	1.0	9.0	3.1	2.6	1.7	10.3*	.8	10.1	3.3	100
MacNeil/Lehrer Report	27.1*	18.2*	24.3*	12.3	—	—	6.8	.2	.3	—	3.1	3.0	1.3	3.5	100
Children's															
Sesame Street	5.2	16.5*	4.0	12.1*	33.2*	4.7	9.3	1.8	2.6	3.3	3.6	1.1	1.2	1.5	100
Mister Rogers	1.0	16.1*	—	9.4*	47.3*	7.2	8.8	.1	1.0	6.3	2.1	—	—	.7	100
Electric Company	5.0	14.4*	4.1	12.0*	33.8*	5.0	8.7	4.0	2.0	4.7	3.4	.5	.8	1.5	100
Other															
Women—talk show	4.9	2.8	7.2	13.2*	8.7	—	3.3	1.0	1.5	—	19.1*	3.0	30.0*	5.3	100
The French Chef	12.6*	17.5*	4.2	20.4*	7.4	1.8	9.3	—	2.3	3.7	10.9	3.9	2.7	3.3	100

*Indicates three largest segment shares for each program.

a. Letters associated with each segment indicate which concentration it is in, namely: AF = Adult Female, AM = Adult Male, Y = Youth, and M = Mixed.

Even among these five there is a marked difference in PTV exposure. The first two are 100% and 92% above average, respectively, while the latter three have PTV Exposure Indices that are considerably lower, ranging from 5% to 20% above average.

The overall program mix of PTV as of the time of the survey, late 1977-1978, has done a far better job of attracting the members of the Arts and Cultural Activities and the Cosmopolitan Self-Enrichment segments than those in any of the remaining fourteen segments. They are the only two segments in which more than half the membership (52.2% and 50.1%, respectively) reported watching PTV once a week or more (Table 4.2b). No other segment is even close. The next highest proportion is 30.2% for people in the News and Information segment. Both of the heaviest PTV viewing segments have a substantially lower percentage of people in the never/undetermined category than do any of the other twelve segments. Their respective never/undetermined proportions are 27.7% and 15.9% compared with a minimum percentage of 46.6% among the other twelve segments and a maximum of 75.9%. One of every three people watching PTV once a week or more in the U.S. audience 13 years of age or older is a member of one of these two segments, although they account for only 16.6% of the population. Furthermore, they are among the three highest audience share segments for a large majority of the 22 PTV programs measured (Table 4.3).

The Arts and Cultural Activities and Cosmopolitan Self-Enrichment segments share the following characteristics:

(1) an interest in abstraction and culturally upscale subject matter (both score exceptionally high on the Classical Arts and Comprehensive News and Information factors);
(2) exceptionally high scores on needs for Intellectual Stimulation and Growth as well as Understanding Others; and
(3) unusually high levels of both education and income.

It comes as no surprise that the two interest segments with this profile are also the two segments whose members watch PTV more than those in any of the other twelve segments. Though these two segments are both heavy users of PTV, their choices of specific PTV programs differ.

With few exceptions (principally children's programs) the Arts and Cultural Activities segment accounts for one of the three largest audience shares of each of the 22 PTV programs reported in Table 4.3. For fourteen of these programs they account for a larger share of audience than do those in any other segment. The leading shares range from a low of 19.8% of the audience for *The Best of Families* to a high of 51.3% for *Opera*. Though they account for only 9% of the entire population, they comprise over a third of the audience for such PTV programs as *Dickens of London, Visions, Upstairs, Downstairs, Great Performances, Evening at Symphony, Opera,* and *Evening at Pops*. Members of this segment clearly have constituted the primary audience for PTV classical arts programming.

Members of the Cosmopolitan Self-Enrichment segment, though similar in their overall PTV viewing to those in the Arts and Cultural Activities segment, nonetheless are attracted to watching a somewhat different menu of PTV programming, namely:

(1) Except for four of the 22 PTV programs, their share of audience is substantially less (often by more than ten percentage points) than the corresponding figures for those in the Arts and Cultural Activities segment.
(2) Three of the four programs for which they have higher audience shares are children's programs.
(3) The other program for which they have a higher share is *Once Upon a Classic* (31.4% versus 17.5%). The only other program for which they account for more than a 20% share of audience is *Dickens of London* (27.1%).
(4) In spite of their broad interests and their high levels of education and PTV viewing, they account for an extremely modest audience share of *Washington Week in Review* (6.0%) and *Wall Street Week* (5.3%). The corresponding figures for the Arts and Cultural Activities segment are 22.6% and 32.6%.
(5) They are a bit more attracted to the *McNeil/Lehrer Report*, accounting for 18.2% of its audience compared to a share of 27.1% for those in the Arts and Cultural Activities segment.

This pattern of differentiation can be accounted for by at least two different factors. First, people in the Cosmopolitan Self-Enrichment segment have younger children than do those in the other segment. The average age of their children is 10 years versus 12 years in the Arts and Cultural Activities segment. Hence, the former group watches more children's programming on PTV.

Second, Cosmopolitan Self-Enrichment segment members are more print-oriented (books, magazines, and newspapers) than are those in any other segment.[2] For example, they are 286% above the average for the population in their frequency of reading books. It is likely that their preference for the use of print media at least partially accounts for their somewhat lower frequency of PTV viewing since both activities compete for some of the same leisure time. This hypothesis is further supported by the fact that the two PTV programs for which they have the highest audience shares have formats that are closest to those of books, namely, *Once Upon a Classic* and *Dickens of London*. The one news program, the *McNeil/Lehrer Report,* for which they have a high audience share, is more booklike than other news programs, in that it has a format that treats a single subject in depth each time it is aired rather than reporting more superficially on a broader range of subject matter. In the absence of this booklike, in-depth analysis, the diet for news for this segment is more apt to be satisfied by print media, and hence the audience shares for the other PTV news programs are quite low.

People in the Cosmopolitan Self-Enrichment segment probably have less of an involvement with television in general even though they are relatively heavy PTV viewers. They tend to be somewhat more selective in their viewing than persons in the Arts and Cultural Activities segment, using PTV to supplement their print media preferences and to assist in the education of their children.

Those in the News and Information segment constitute the third heaviest viewing group, with a PTV Exposure Index 20% above average. Nevertheless, only 30.2% of them report usually watching PTV once a week or more. Almost half the people in this segment (49.6%) are in the never/undetermined category. As one would expect, given their predominant interest in being informed on a wide range of subject matter, their PTV viewing is concentrated more on programs high in informational content rather than on those emphasizing musical or literary artistry.

There are nine programs for which members of this segment rank among the top three segments in audience share. Seven of these are predominantly informational in character, namely:

- *Age of Uncertainty*
- *Visions*
- *In Pursuit of Liberty*
- *Washington Week in Review*
- *Wall Street Week*
- *Black Perspective on the News*
- *McNeil/Lehrer Report*

Not only is their focus of viewing much narrower than for people in the two segments discussed previously, but the benefits they appear to seek also differ. Individuals in the Arts and Cultural Activities and the Cosmopolitan Self-Enrichment segments have high needs for intellectual stimulation and make more use of programming that emphasizes abstraction and intellectually upscaled content. In contrast, those in the News and Information segment have needs that are more concerned with being socially stimulating and maintaining family ties than with intellectual stimulation. Their watching of PTV serves their socialization needs.

In discussing people in the Highly Diversified segment in the previous chapter, we concluded that, although they lacked the educational and social advantages of others (e.g., especially those in the Arts and Cultural Activities and Cosmopolitan Self-Enrichment segments), they nonetheless were seeking to expand their intellectual and cultural horizons.

This conclusion is consistent with their relatively high score on PTV viewing. Their PTV Exposure Index is 115%. They have virtually the same proportions of frequent and never/undetermined PTV viewers as do those in the News and Information segment (29.4% and 51.4%, respectively). They are above-average viewers for 16 out of the 22 PTV programs in that they account for a higher audience share than their relative numbers in the population (i.e., 8% of the audience) would indicate. They are among the top three segments in audience shares for 10 of the 22 PTV programs. Of the ten, they account for a high of 27.0% of the audience for *Black Perspective on the News*. The segment is one-third Black, which is a higher percentage of Blacks than in any other segment and triple that of the population in general. They are among the top three segments for all three children's programs, as well as such programs as *Great Performances, In Pursuit of Liberty, The Best of Families, Nova,* and *Opera*.

This group is attracted to intellectually upscale material that is both adult- and child-oriented. PTV is apparently seen as appropriate for both themselves and their children. Their exceptionally broad range of interests probably accounts, in part, for this diversity of viewing behavior. In addition, as will be developed in more detail in Chapter 5, viewing in these households appears to be more of a family affair than in the other segments. The entire household participates in choosing what commercial and public television programs will be viewed, and they alternate their program selections from time to time to fit the preferences of different family members.

The membership of the Family-Integrated Activities segment is only slightly above average on their score for the PTV Exposure Index (105% of average). One of every four members of this segment reports watching PTV once a week or more, with 46.6% in the never/undetermined category.

They watch PTV children's programs more than people in any other segment. They account for 47.3% of the audience for *Mister Rogers,* 33.8% of that for *Electric Company,* and 33.2% of that for *Sesame Street*. They also are among the top three segments in their exposure to *Nova, Masterpiece Theatre,* and *Once Upon a Classic*. Their high score on children's programs is to be expected. Members of this segment are 87% female, 75% are adults with children, and the average age of their children is nine. They are more apt to have children and the children are younger than in any other segment.

A detailed analysis of their television viewing over all programs (both commercial and public) led us to conclude that, in addition to their presence, children have more of an impact on the adult members of the segment than do those in any other segment.[3] The television programs selected by people in this segment have a broad range of appeal across age groups and appear to be used as a means of bringing adults and children together to share the viewing experience. Besides sharing, there is evidence (e.g., the children's programs viewed) of their directing their children's viewing to program alternatives that contribute primarily to their children's development and, for some of their choices, to their own as well.[4]

This sharing of experience, combined with guidance, places individuals in this segment somewhat apart from those in the Highly Diversified segment. The latter segment, though also oriented toward family viewing, shows less evidence of consistently guiding their children toward programs that contribute to personal development.

Though the above segments are similar in that they are above-average viewers of PTV, each represents a relatively unique mix of needs and interests, which in turn is associated with differences in both the amount of their PTV viewing behavior and its content.

Below-Average Viewing Segments

There are nine interest segments with below-average PTV Exposure Indices. Starting with persons in the least exposed segments and ending with those in the most, they are as follows:

- Family- and Community-Centered (Adult Male Concentration)—a PTV Exposure Index of 48% of average
- Elderly Concerns (Adult Female Concentration)—50%
- Money and Nature's Products (Adult Male Concentration)—57%
- Detached (Mixed)—61%
- Competitive Sports and Science/Engineering (Youth Concentration)—66%
- Indoor Games and Social Activities (Youth Concentration)—79%
- Mechanics and Outdoor Life (Adult Male Concentration)—83%
- Home- and Community-Centered (Adult Female Concentration)—88%
- Athletic and Social Activities (Youth Concentration)—97%

Members in a number of these below-average PTV viewing segments make so little use of PTV that data about their actual viewing provide little insight into their behavior. This is in contrast to those in the above-average viewing segments, where differences in members' program preferences were useful, together with knowledge of their interests, needs, and demographics, in developing our conclusions. In effect, at this point one is faced with a puzzle with two few pieces available to see the pattern. Therefore, we have chosen to briefly describe the PTV viewing characteristics of each segment. Some conclusions regarding their PTV behavior are also offered. For now, there is virtually no discussion as to each segment's interests, needs, demographics, and media usage in support of these conclusions. The rationale is presented in Chapter 6, where other media usage data are also available. At that point the puzzle will be sufficiently complete, so that even in the case of the below-average PTV viewing segments, the basis for our conclusions will be apparent.[5] Our thumbnail sketches of each segment's PTV usage are as follows:

(1) Family- and Community-Centered. Only 12.3% of this segment's members report frequently watching PTV, while almost two-thirds either are not sure or never have watched PTV. They do not rank among the top three segments in their audience share of any of the 22 PTV programs. Their highest scoring PTV programs are *Upstairs, Downstairs* (9.0%), *Washington Week in Review* (6.2%), and *Wall Street Week* (7.2%). Though these people have broad interests and are somewhat above average in education, they are below-average users of all media except newspapers, especially local papers. For those in this segment, family ties score higher as a need than for any other segment. It is likely that their low usage of PTV, as well as other media, is due to their participation in other family and community activities that compete for time. They score high only on those media that provide information useful for planning these types of activities. A second factor that tends to reduce PTV usage more than that of other media is that they are not especially attracted to abstract

and culturally upscaled material such as that which much of PTV comprises.

(2) Elderly Concerns. As in the case of the preceding segment, few members (12.5%) of this segment usually watch PTV once a week or more. Three-quarters (75.9%) are in the never/undetermined category. In spite of this they accounted for a higher proportion of the audience (30.0%) than did any other segment for one PTV program, namely, *Women,* a talk show. For one other program, *Upstairs, Downstairs* (10.8%), they were among the top three segments. These programs help to cope with their needs for vicarious participation, social integration, and acceptance. Relatively few PTV programs are structured in a fashion that successfully addresses this pattern of needs.

(3) Money and Nature's Products. The overall PTV viewing levels for members of this segment are virtually the same as those for the previous two. PTV coverage for persons in this segment is quite modest. There are, however, three programs for which they are among the top three segments in audience share, *Washington Week in Review* (16.6%), *Wall Street Week* (9.6%), and *Evening at Symphony* (7.5%). For the latter two programs their audience share is less than half that of the next highest segment, while for *Washington Week in Review* these people rank second with a share of audience close to those of both the other top-ranking segments (22.6% and 14.3%). Their interest in watching financial/business programming may in part be related to their overall interest in activities that yield some form of tangible return. Their interest in culturally upscale or abstract material is generally quite low. *Evening at Symphony* is more apt to be used as background music for other activities than to focus on symphonic music in and of itself.

(4) Detached. Only 15.6% of this segment's member report usually watching PTV one or more times a week and fully 70.9% are in the never/undetermined category. They do, however, rank among the top three segments in audience share for two PTV programs, *Black Perspective on the News* (10.3%) and *Women*

(19.1%). This segment has the third highest percentage of Blacks. People in it make relatively little use of all media, and much of what they do use appears to help them escape from boredom as well as problems.

(5) Competitive Sports and Science/Engineering. Among the people in this segment, 17.3% report watching PTV frequently and 56.6% are in the never/undetermined category. For one program, *Age of Uncertainty,* this segment's share of 16.8% ranks it among the top three segments. The next highest scoring programs based on audience share are *Mister Rogers* (6.3%), *Nova* (5.7%), and *Electric Company* (4.7%). These adolescent males are below average in their interest in abstract and culturally upscale subject matter, as are most members of the other two youth-oriented segments (Athletic and Social Activities and Indoor Games and Social Activities). Their viewing of *Nova* and *Age of Uncertainty* may be a result of school homework assignments rather than their voluntary action. Their somewhat higher scores on children's programs most likely result from the presence of younger children and only to a minor degree reflect their own preferences.

(6) Indoor Games and Social Activities. One of five members of this segment watches PTV frequently with almost two-thirds in the never/undetermined category. Individuals in this segment are below average in their viewing of all 22 PTV programs. The only program for which they account for more than 5% of the audience is *The Best of Families* (5.2%). These people have relatively little interest in intellectually upscale or abstract subject matter. Their strongest needs are primarily related to peer-group socialization, namely, a need for status and to be socially stimulating. They do watch a lot of commercial television. These young, lower socioeconomic status women, many of whom have started families, use television as a major means of entertainment. They are more interested in *Soul Train* and *American Bandstand* than in *Evening at Symphony* or *Opera* and more interested in *Charlie's*

Angels and *Starsky and Hutch* than in *Once Upon a Classic* or *In Pursuit of Liberty.*

(7) Mechanics and Outdoor Life. Over half (55.1%) the people in this segment are in the never/undetermined category, while 21.5% watch frequently. They do not rank among the top three segments on any program. The three PTV programs for which they score highest are *Wall Street Week* (7.6%), *Evening at Symphony* (5.7%), and *Evening at Pops* (4.6%). Their general media usage level across all media is at or below average. They are not especially attracted to abstract or intellectually upscale material. They have high needs to escape. These people are interested in outdoor activities that are individualistic, noncompetitive, and that emphasize personal physical accomplishment and manual dexterity. Their usage is high for media such as "How to" books and automotive magazines that cover the types of activities in which they are interested. Similarly, when television programming picks up substantial elements of escape, where the forces of good and evil are clearly drawn, as is the case for some science fiction and drama, their viewing levels tend to increase.

(8) Home- and Community-Centered. The overall report of PTV viewing for persons in this segment is modest (close in magnitude to the previous segment). They score among the top three segments for only 1 of the 22 programs, namely, *Dickens of London* (6.5%). Nonetheless, there are eight programs for which they account for more than their relative size of the total population (8%), and therefore are above-average viewers. Their share of audience for these eight programs ranges from 8.7% to 10.3%. They are: *Visions, In Pursuit of Liberty, Washington Week in Review, Black Perspective on the News, Sesame Street, Mister Rogers, Electric Company,* and *The French Chef.* These predominantly adult, female homemakers use almost all types of media at or near the average for the population as a whole. Their interests, as the segment name implies, are oriented to their homes and communities. They do not have extremely high or low scores

on any of their needs. Though some people in this segment are apt to be a part of the audience for many types of media, they are not likely to be the dominant segment for any.

(9) Athletic and Social Activities. Members of this segment are only slightly below average in their PTV Exposure Index (97%). Nevertheless, they report never watching 8 of the 22 PTV programs. They account for 5% or more of the audience of only two PTV programs, *Mister Rogers* (7.2%) and *Electric Company* (5.0%). In common with other Youth Concentration segments, this group of predominantly young women is not attracted to watching or reading information-oriented, abstract, culturally upscale subject matter. Their audience shares are especially low because this segment constitutes only 4.2% of the population.

Conclusions

As measured by the differences in the extent to which members of the fourteen interest segments choose to watch PTV:

PTV has had an extremely uneven appeal across the fourteen interest segments.

(1) At one extreme, about half the members of two segments watch PTV once a week or more. At the other extreme, fewer than 13% do.
(2) At one extreme, fewer than 28% of the segment members report either never watching or not being able to recall watching PTV. At the other extreme, about two-thirds of the segment's members report the same lack of familiarity with PTV.

Among people in the five above-average viewing segments PTV serves a multiplicity of functions which, in turn, are associated with different patterns of PTV program choices. For example:

(1) Cosmopolitan Self-Enrichment segment members watch less frequently than those in the Arts and Cultural Activities segment except for those programs that are booklike. For such programs they watch more. Though both segments are attracted to abstract, intellectually upscale material, it appears that individuals in the Cosmopolitan Self-Enrichment segment are

primarily attracted to booklike print media, and are exceptionally heavy viewers only of PTV programming that has booklike characteristics.

(2) People in the News and Information and Family-Integrated Activities segments are primarily attracted to PTV for a more narrow range of reasons. For those in the News and Information segment the value of informational programming, such as news, is not its intellectual content so much as its usefulness as a catalyst to socializing. For those in the Family-Integrated Activities segment the predominant pattern of PTV program choices focuses on their children's personal development, rather than on their own self-development.

The limited attraction of PTV to people in the below-average segments also occurs for a multiplicity of reasons. This can be seen even without detailed data on their usage of other media, discussed in the chapter that follows. For example:

(1) For individuals in the Elderly Concerns segment, the need for social support and help in coping with problems of loneliness is important. The PTV programming included in our analysis does not effectively address these needs.

(2) For those in the Family- and Community-Centered segment, activities associated with family and community life take precedence over time spent being exposed to PTV, and to other media as well. The most attractive media to this segment, in terms of relative usage, are those that provide useful information related to these activities.

(3) Competitive Sports and Science/Engineering segment members are not attracted to culturally upscale or abstract material. Their modest viewing of PTV seems to occur when it is forced upon them by homework assignments or the presence of younger family members.

PTV is relatively weakest among the interest segments whose membership is predominantly young, male, or both.

(1) None of the three Youth Concentration segments is above average in its PTV Exposure Index, although people in the Athletic and Social Activities segment are close to average (97% of average).

(2) None of the three Adult Male Concentration segments is above average. Two are among the lowest three of the fourteen segments in their Exposure Indices with scores of 57% and 48%.

Some elements of PTV programming diversity have had more of an impact on attracting diverse audiences than others.

(1) Though *Dickens of London, Upstairs, Downstairs,* and *Great Performances* have diverse content, they tend to draw more than half of their respective audiences from the same two interest segments. These are the Arts and Cultural Activities and Cosmopolitan Self-Enrichment segments.

(2) Children's programming, on the other hand, obtains its largest audience share from people in the Family-Integrated Activities segment and, in turn, obtains less than 5.2% of its audience from members of the Arts and Cultural Activities segment. These results, of course, do not reflect the viewing of children under age 13.

(3) One-half the viewing audience of *Women* comes from people in the Elderly Concerns, Detached, and Highly Diversified segments.

If PTV is to serve all of its audiences, a goal that both its Board of Trustees and the two Carnegie Commissions have endorsed, it inevitably must recognize and respond to, in its programming and promotional strategies, the fact that the audience it wishes to attract is pluralistic. It is a highly diverse audience with varied interests and needs that viewers seek to satisfy through their media behavior.

Given the findings and conclusions reported above, it is obvious that attracting more viewers from each of the various segments will require that PTV substantially increase its variety of substantive content, and program execution (e.g., documentaries versus drama), as well as enhance the nature of the advertising and promotional activities used to inform potential viewers of the availability of PTV programs.

In the next two chapters we discuss the use of other media by people in the above- and below-average PTV segments. By understanding their responses to the competition for leisure time we can gain additional insights into opportunities that might exist for PTV to be responsive to people's interests and needs. In the last chapter we pull together our conclusions from this and subsequent chapters to discuss the usefulness of this information for helping those concerned with the future of PTV.

Notes

1. For detailed PTV program coverage and usage data for each of the 22 PTV programs included in the study, see the Appendix, Tables A-1 and A-2.

2. Findings in support of this conclusion are reported in Chapter 5 of this book as well as in Frank and Greenberg (1980: ch. 7).

3. See Frank and Greenberg (1980: ch. 5).

4. Those readers familiar with our previous book (Frank and Greenberg, 1980) will recognize that the conclusions regarding the nature of each segment's degree of attraction to various subject matters are the same as those appearing in this section and the next chapter. What differs is that in this book we are concerned solely with understanding PTV behavior, and not with the broader development and validation of our interest segmentation scheme, or even understanding television usage in general.

5. Media such as commercial television, books, magazines, movies, newspapers, and radio are among the major PTV competitors for the time of the people in each segment. Knowing how those who watch little PTV spend their time can, therefore, be extremely helpful in understanding why PTV receives a low priority for them and what types of PTV programming might attract them. It may also help to identify means of using other media to advertise PTV programming targeted against these eight viewing segments.

5

Media Usage Among Above-Average PTV User Segments

In this and the next chapter, each segment's use of PTV is placed in the broader context of members' usage of other media, namely, commercial television, books, magazines, movies, newspapers, and radio.

Background Details

Table 5.1 reports the pattern of overall usage by interest segment for books, television, magazines, movies, newspapers, and radio.[1] Though this chapter discusses our findings for only the five segments with above-average viewing of PTV, both Tables 5.1 and 5.2 include data for all fourteen segments. These two tables will be used as the basis for our discussion in Chapter 6 of the segments with below-average viewing of PTV. The figures in the body of the table are percentaged ratios of each segment's usage of a given medium to that for the entire population. For example, members of the Cosmopolitan Self-Enrichment segment had a reported annual frequency of movie attendance 28% greater (therefore, a ratio of 128) than the average number of movies attended for the entire population, which is six per year.

Though these usage ratios are computed in the same manner across all media, the absolute measures of usage vary. For each medium, the measures used, together with their average for the entire population, are as follows:

(1) *public television*—percentage of people who watched once a week or more (26%)

(2) *television*—average number of hours during which some viewing occurred during a typical week, including both commercial and public stations (36 hours)
(3) *books*—average number read in past year (16 books)
(4) *magazines*—average number of magazines read "regularly" (5 magazines)
(5) *movies*—average number of times gone to movie in past year (6 times)
(6) *daily newspapers*—average number of days per week read (4 days)
(7) *financial newspapers*—percentage of people who read at least one (*Barrons* or *Wall Street Journal*) regularly (4%)
(8) *local weeklies*—percentage of people who usually read (48%)
(9) *Sunday newspapers*—percentage of people who usually read (71%)
(10) *Sunday supplements*—percentage of people who read at least one regularly (35%)
(11) *radio*—average number of hours during which some listening occurs in a typical week (18 hours)

Table 5.2 reports the average frequency of viewing each of nineteen television program types by interest segment. These data are computed from respondents' ratings of the extent of their viewing each of 149 programs or program types. For each of the 149 programs, respondents were asked if they ever watched it and, if so, how often during the last four weeks. These 149 ratings were later summarized into the 19 program categories listed in Table 5.2. The programs, both commercial and public, within each type are listed in Figure 5.1.

We recognize that some of these program types are not as homogeneous as one would like them to be and that some real anomalies exist, but we also feel constrained for conceptual reasons to limit the number of categories as much as possible. Certainly, *Fernwood 2 Night* is different from other programs in the talk show category and might, indeed, be sufficiently unique to constitute a category of its own. The variety of types of music in the Musical Performances category is clearly great. However, rather than allow our preconceived notions about the types of audiences that would be attracted to a program to influence our categorization, we elected to use the system outlined in Figure 5.1

(text continues on p. 108)

Table 5.1 Media Usage by Type of Media and Interest Segment[a]

Type of Media	Entire Population (average)	Arts and Cultural Activities (AF)[b]	Cosmopolitan Self-Enrichment (M)	News and Information (M)	Highly Diversified (M)	Family-Integrated Activities (AF)	Athletic and Social Activities (Y)	Home- and Community-Centered (AF)	Mechanics and Outdoor Life (AM)	Indoor Games and Social Activities (Y)	Competitive Sports and Science/Engineering (Y)	Detached (M)	Money and Nature's Products (AM)	Elderly Concerns (AF)	Family- and Community-Centered (AM)
Television															
Public	26%	200*	192*	120*	115	105	97	88	83	79	67	61	57	50	48
Overall	36 hours	97	75	134*	120*	106	80	92	89	114	101	93	98	118*	87
Books	16 books	154*	238*	99	83	92	67	90	90	162*	103	49	30	70	69
Magazines	5 magazines	114	134*	96	150*	132*	98	100	92	108	106	52	72	56	80
Movies	6 movies	84	128	70	139	102	187*	98	149*	131	172*	67	38	18	48
Newspapers															
Daily	4 days	120*	110	117*	100	98	63	110	80	63	95	88	110	98	115*
Financial	4%	114	267*	171*	126*	93	12	83	64	74	33	12	69	0	114
Local Weekly	48%	102	104	114	121*	102	66	116*	92	85	93	74	94	103	115*
Sunday															
Paper	71%	117*	116*	127*	92	98	70	114	75	68	108	95	100	95	107
Supplements	35%	132*	178*	132*	98	140*	60	93	62	83	79	57	72	69	112
Radio	18 hours	74	125*	78	120	122	161*	91	134*	79	95	93	69	69	72

*Indicates three highest segment scores for each type of media.
a. Figures are percentaged ratios of segment's usage of a given medium to that for the entire population.
b. Letters associated with each segment indicate which concentration it is in, namely: AF = Adult Female, AM = Adult Male, Y = Youth, and M = Mixed.

Table 5.2 Viewing Frequency Ratios by Program Type and Interest Segment[a]

Type of Program	Entire Population (average)	Arts and Cultural Activities (AF)[b]	Cosmopolitan Self-Enrichment (M)	News and Information (M)	Highly Diversified (M)	Family-Integrated Activities (AF)	Athletic and Social Activities (Y)	Home- and Community-Centered (AF)	Mechanics and Outdoor Life (AM)	Indoor Games and Social Activities (Y)	Competitive Sports and Science/Engineering (Y)	Detached (M)	Money and Nature's Products (AM)	Elderly Concerns (AF)	Family- and Community-Centered (AM)
Television Exposure															
Public	26%	200*	192*	120*	115	105	97	88	83	79	67	61	57	50	48
Overall	36 (hrs/wk)	97	75	134*	120	106	80	92	89	114	101	93	98	118*	87
Adventures	4.14	56	42	102	164*	107	90	91	124*	145*	109	85	104	111	92
Children's Programs (3 of 6)[c]	1.36	46	103	81	165*	224*	121	111	82	132*	119	71	46	34	46
Crime Dramas	7.70	81	51	118*	139*	100	76	96	109	139*	105	97	109	109	85
Documentaries (2 of 3)	.34	171*	129*	132*	126	126	3	91	76	44	94	59	129*	88	59
Dramas	4.42	81	66	126*	119	129*	87	113	81	105	72	83	109	134*	90
Game Shows	3.56	89	48	141*	123	123	58	121	60	166*	103	79	69	169*	62
Movies	5.45	84	79	133*	147*	123*	108	84	108	90	97	93	92	81	77
Musical Performances (3 of 6)	1.02	192*	94	106	211*	77	91	90	52	162*	88	72	57	75	30
News/Commentaries (4 of 8)	2.37	157*	102	233*	112	64	35	84	58	27	48	66	160*	133	113
News Shows–Daily	5.19	127*	134*	134*	103	96	40	99	70	46	71	80	113	126	117
Science Fiction	1.25	70	71	94	174*	86	122	80	127	184*	131*	93	62	75	74

Situation Comedies	19.15	87	30	121*	118	103	115	114	93	149*	130*	82	77	105	63
Soap Operas	3.64	76	26	140	127	146	86	180*	25	154*	29	99	77	196*	33
Specials	.89	153*	117	118	137*	112	64	113	85	96	65	39	119*	81	71
Sports	5.79	101	86	167*	121	51	69	79	72	86	205*	73	132*	73	131
Talk Shows (1 of 11)	2.90	143*	127	182*	92	96	42	119	29	76	76	78	113	139*	68
Theatrical Performances (all 8)	.63	386*	216*	124*	100	70	19	46	29	60	19	46	68	73	56
Variety Shows	3.63	86	72	154*	133*	90	79	108	79	131	83	73	117	140*	28
Others (1 of 3)	.67	94	31	154	164*	48	64	164*	13	64	39	100	136	191*	133

*Indicates three highest segment scores for each type of program.

a. Percentaged ratio of segment's viewing frequency of a given program to that for the entire population. The statistical significance of each of the nineteen program type measures contained in this table was evaluated based on univariate F ratios with 13 and 2462 degrees of freedom. All nineteen F ratios are significant at the .005 level.

b. Letters associated with each segment indicate which concentration it is in, namely: AF = Adult Female, AM = Adult Male, Y = Youth, and M = Mixed.

c. Parentheses only occur after program types that include one or more PTV programs. The counts given are the number of PTV programs out of the total number in the category.

Figure 5.1
Television Programs by Type

Adventures
    Nancy Drew and the Hardy Boys
    The Wonderful World of Disney
    Six Million Dollar Man
    Young Dan'l Boone
    The Life and Times of Grizzly Adams
    The Bionic Woman

Children's Programs
    Captain Kangaroo
    Captain Noah
    *Sesame Street
    *Mister Rogers
    *Electric Company
    Children's Cartoons

Crime Dramas
    Kojak
    Police Woman
    Charlie's Angels
    Baretta
    Chips
    Hawaii Five-O
    Barnaby Jones
    Rosetti and Ryan
    Switch
    The Rockford Files
    Quincy
    Starsky and Hutch

Documentaries
    *Age of Uncertainty
    *Nova
    Last of the Wild

Dramas
    Little House on the Prairie
    Rafferty
    The Fitzpatricks
    Lou Grant
    Family
    Oregon Trail
    Big Hawaii
    The Waltons

Game Shows
    Hollywood Squares
    The Price Is Right
    Wheel of Fortune
    Family Feud
    It's Anybody's Guess
    Gong Show
    Tattletales
    20,000 Dollar Pyramid
    Match Game '77

Movies
    Sunday Night
    The Big Event
    Monday Night
    Wednesday Night
    Friday Night
    Saturday Night
    Late Night

Musical Performances
    *Evening at Symphony
    *Opera
    *Evening at Pops
    American Bandstand
    Soul Train
    Music Hall America

News/Commentaries
    Sixty Minutes
    *Washington Week in Review
    *Wall Street Week
    *Black Perspective on the News
    Evening Magazine
    *MacNeil/Lehrer Report
    Face the Nation
    Meet the Press

News Shows Daily
    Local News
    National Network News
    (ABC, CBS, NBC)

(continued)

ABOVE-AVERAGE USER SEGMENTS

Figure 5.1 Continued

Science Fiction
- The Man from Atlantis
- The New Adventures of Wonder Woman
- Logan's Run

Situation Comedies
- Rhoda
- On Own Our
- All in the Family
- Alice
- The San Pedro Beach Bums
- The Betty White Show
- Maude
- Happy Days
- Laverne and Shirley
- Three's Company
- Soap
- M*A*S*H
- One Day at a Time
- Mulligan's Stew
- Eight Is Enough
- Good Times
- Bustin Loose
- Welcome Back, Kotter
- What's Happening
- Barney Miller
- Carter Country
- Sanford and Sons
- Chico and the Man
- Bob Newhart
- We've Got Each Other
- Fish
- Operation Petticoat
- The Jeffersons
- Tony Randall
- The Love Boat

Soap Operas
- Love of Life
- The Young and the Restless
- Search for Tomorrow
- All My Children
- Days of Our Lives
- As the World Turns
- The Doctors
- One Life to Live
- The Guiding Light
- Another World
- General Hospital

Specials
- NFL Football
- Wide World of Sports
- Football
- Basketball
- Hockey
- Baseball
- Other

Talk Shows
- Fernwood 2 Night
- Johnny Carson
- Today
- Good Morning America
- Phil Donahue
- Joel A. Spivak
- Dialing for Dollars
- *Women
- Mike Douglas
- Dinah
- Merv Griffin Show

Theatrical Performances
- *Dickens of London
- *Visions
- *Upstairs, Downstairs
- *In Pursuit of Liberty
- *Great Performances
- *The Best of Families
- *Masterpiece Theatre
- *Once Upon a Classic

Variety Shows
- The Richard Pryor Show
- Redd Foxx
- Donnie and Marie
- The Carol Burnett Show
- Lawrence Welk
- Saturday Night Live
- Andy Williams

Others
- *The French Chef
- Any Religious Programs
- Any Spanish Programs

*PTV programs.

and to account for apparent anomalies in segment viewing in the course of the discussion if and when they emerged.

The first column of Table 5.2 contains the average viewing frequency by program for the population. For example, situation comedies were watched on an average of 19.15 times during the preceding four weeks, while documentaries were watched on average only one-third of a time per week (i.e., approximately once every three weeks).

In order to facilitate comparing the relative frequency of viewing of each program type across segments, the body of Table 5.2 contains the percentaged ratio of each segment's score for the program type to the corresponding average for the population as a whole. These ratios are to be interpreted in the same fashion as those in Table 5.1.

The numbers in parentheses following some of the program type labels report the number of PTV programs and total programs contained in that type. For example, three of the six children's programs are aired on PTV. Program type labels not followed by numbers in parentheses contain programs aired solely on commercial television.

The variation in viewing frequencies across all segments for each of the nineteen program types has been tested for statistical significance using the univariate F-ratio statistic. All of the nineteen tests were significant at the .005 level.

In the discussion of findings it is often useful to include more detailed findings on each medium (e.g., findings related to magazine types and individual magazines) than are contained in Tables 5.1 and 5.2. These more detailed data are not included in this book as they are quite voluminous and have been reported elsewhere (Frank and Greenberg, 1980). Readers interested in page references to these data should see Note 1 at the end of this chapter.

Interest Segment Findings

Above-average users of PTV are, for the most part, also above-average users of other media. For each of the ten media in Table

5.1 (excluding PTV) the ratios for those segments with the three highest scores are indicated with asterisks. In the case of ties within a given medium, more than three segments are asterisked. Of the 31 top scores, 21 occur in the 5 segments whose members are above-average users of PTV.

At the other extreme, as one would expect, below-average users of PTV tend to be below-average users of other media. Though they are not marked in Table 5.1, of the 30 lowest scores all 30 occur for segments with below-average usage of PTV.

People in segments with above-average PTV usage make more extensive use of a broader range of types of media than do those in segments having below-average PTV usage. Only one of the ten media completely departs from this pattern, namely, movie attendance. The three segments with the highest levels of movie attendance exhibit below-average PTV usage.

Within the context of this overall pattern of media usage, there are differentiating patterns of media usage among the segments comprising the five above-average PTV usage categories.

Arts and Cultural Activities

Highly educated, adult women in households with manager or professional as head. Broad range of intellectual and cultural interests—especially classical arts. Low interest in household activities and management. High needs for intellectual stimulation and growth and for understanding others with low needs for status enhancement and escape.

The people in this segment make use of a broad range of media. They are above the average for the entire population in their use of nine of the eleven media in Table 5.1. They are among the three segments with the highest usage levels for five media.

The two media on which they score below average are radio and overall television, on which they score 74% and 97% of average, respectively. Their slightly below average score for overall television is not inconsistent with their high PTV score, as the overall television measure is dominated by commercial rather than public viewing.

The television program types for which members of this segment rank among the top three segments with respect to viewing are of two kinds. The first are those program types in which all, or at least half, of the programs are aired on PTV. The second are programs that tend to attract people with unusually high educational levels who score high on needs for Intellectual Stimulation and Growth.

These people are relatively heavy viewers of theatrical performances, documentaries, musical performances, and news/commentary programs, all of which include shows of which at least half are aired on PTV. They are also well above average in their viewing of the talk show category which, in addition to *Women* shown on PTV, includes *The Today Show*, *The Phil Donahue Show*, and *Dinah*. Their high score on specials results primarily from exceptionally heavy viewing of *Washington Behind Closed Doors*, a drama about political intrigue that was aired in the fall of 1977. This segment is also among the top three in their viewing of local and network news programs.

The content to which they are attracted in media other than television is also consistent with their patterns of interests and needs. For example, they are relatively heavy readers of historical novels and books on topical areas, such as music, opera, and dance. Among all fourteen segments, they are the heaviest regular readers of news and information sections of the newspaper and the second heaviest readers of the entertainment section. These people are among the most frequent regular readers of three magazine types, select (e.g., *National Geographic* and *Psychology Today*), business/finance, and news. Finally, despite their below-average use of radio, they report the highest frequency of listening to classical music and instrumental "background music" and are among the heaviest listeners of news and educational programming.

The people in this Arts and Cultural Activities segment consistently use all eleven types of media in a selective fashion to broaden their intellectual and cultural horizons and to satisfy their informational needs. Their viewing of PTV clearly reflects a more general lifestyle that satisfies their needs for intellectual stimulation and growth by emphasizing interests in abstract, culturally upscale subject matter. For them the absorption of culture and information simultaneously offers both knowledge and entertainment.

Cosmopolitan Self-Enrichment

> Extremely high socioeconomic profile. Diverse pattern of intellectual and cultural interests. Physically active. High needs for intellectual stimulation, unique/creative accomplishment and understanding others. Low needs for status enhancement and for escape from boredom.

This segment's membership is composed of above-average users of ten of the eleven media. Their indices for books (238% of average) and financial newspapers (267%) are substantially higher than any of the other 152 scores for all of the other media/segment combinations reported in Table 5.1. Also extremely high is their readership of Sunday supplements, which is 178% of average, and, as discussed in Chapter 4, PTV vewing (192% of average). The one exception to these above-average scores is their overall television index, which is 75% of average, lower than that for any other segment. This low score results from their relatively infrequent viewing of commercial television.

Though their overall exposure to television is lower than that for people in any of the other segments, they nonetheless rank among the three highest scoring segments for three program types: documentaries, daily news shows, and theatrical performances. Of the three, their score on theatrical performances, which consists entirely of PTV programs, is substantially greater (116% compared to 29% and 34% above average). As previously discussed, their viewing consists, in large part, of book-related programs such as *Once Upon a Classic* and *Dickens of London*.

Members of this segment are also heavy readers of both fiction and nonfiction books, although their scores for nonfiction are a bit higher relative to the entire population than are those for fiction. Their movie attendance, though not high enough to rank them in the top three segments, is still 28% above average and includes material that also tends to be high in intellectual content such as biogrpahies, documentaries, historical or adventure films, and music, opera, and dance films. When it comes to magazines, they score high on five categories: select, news, business/finance, sports, and miscellaneous. They use newspapers as a source of information about entertainment, travel, news, and business. Their use of radio shows an appreciation of quite diverse formats from classical to rhythm and blues.

This segment consists of people who are highly literate, likely intellectuals in the truest sense of the word. They are selective in their exposure to all the mass media, choosing only material that they feel will be intellectually stimulating or informative. In general, they have a bias favoring print media. They become relatively heavy users of nonprint media, such as movies and PTV, only when their content coincides with booklike material that is intellectually rich in content.

News and Information

> Passive interests related to keeping informed on a broad range of subjects and activities. Needs are focused on being socially stimulating and maintaining family ties.

These people report watching television more often than do those in any other segment. They are above average in their viewing of seventeen of the nineteen program types listed in Table 5.2. They rank first or second in their viewing of seven types as shown below:

Rank First
- news/commentary— 133% above average
- talk shows—82%
- variety shows—54%
- news shows/daily—34%

Rank Second
- sports—67%
- movies—33%
- documentary—32% above average

Their high scores for news/commentary and documentaries are in large part due to their heavy viewing of most of the PTV programs contained in those categories. However, the remaining program types on which they rank first or second are composed entirely of commercial programs. Their pattern of both commercial and PTV viewing is quiet consistent.

Their desire to keep informed on a broad range of subjects pervades at least five, if not all seven, top-ranked program types. The only two types for which this interpretation is not obvious are movies and variety shows, although both represent vehicles for keeping up-to-date on what is happening in the entertainment world, an integral part of our social structure.

The desire to keep generally informed also pervades their use of other media. For example:

(1) *Magazines.* The only types on which these people rank in the top three segments are news and general. Within the news category they rank first in their readership of *Newsweek* and *U.S. News and World Report*. Their high rank on general magazines is due to the fact that they rank second on *Reader's Digest* and *TV Guide*.

(2) *Newspapers.* In addition to an overall pattern of above-average usage, they rank first in their readership of all but one of the news-related newspaper content categories included in our study, namely: world, national, local, editorial, and business. The one exception is social news, on which they ranked fifth.

Members of the News and Information segment clearly use media both as an entertainment medium and as a means of keeping informed about the society in which they live. As indicated in Chapters 3 and 4, they do not appear to be social or political activists, nor do they appear to have interests that are driven by a desire for abstraction or culturally upscale material, nor is a broad need for information reflected in their demographic or occupational characteristics. Rather, they appear to seek knowledge about a wide range of subject matter in order to make themselves more socially stimulating and better able to converse with others.

Highly Diversified

Southern, Black, adults with children. Broad range of interests, especially those permitting personal participation with family and/or other informal small group settings. High need for intellectual stimulation and growth.

These individuals are 20% above average in their overall television viewing and 15% above average with respect to PTV viewing. They are at or above average in their viewing of eighteen of the nineteen program types. They rank first or second in their viewing of eight types as shown below:

Rank First
- musical performances— 111% *above* average
- adventure—64%
- movies—47%
- crime drama—39%

Rank Second
- science fiction—74% *above* average
- children's programs—65%
- others (religious)—64%
- specials—37%

As discussed in Chapter 4, in the categories of theatrical performances, musical performances, and children's programs, they are among the top three segments in their viewing of seven PTV programs, namely: *Great Performances, In Pursuit of Liberty, The Best of Families, Opera, Mister Rogers, Electric Company,* and *Sesame Street*. They are also among the top three segments in their viewing of *Nova*. In addition, they rank first in viewing *Black Perspective on the News*. One-third of the people in this segment are Black, a higher percentage than that in any other segment and about three times the percentage in the U.S. population.

Certainly, their television sets play a major role in their family lives. The mixture of heavily viewed programs includes those typically associated with children and adults, males and females, entertainment and education, and both "highbrow" and "lowbrow" material.

The same diversity is reflected in the magazine reading habits of this segment. Their overall index for magazine readership is 50% above average, resulting primarily from their broad readership of many types of magazines. They rank sixth or higher on all fifteen of the magazine content categories included in our analysis; however, they rank first on only one category, Black magazines.

Their radio usage also reflects a broad range of interests. This segment's members rank among the top four segments on all nineteen of the radio content categories.

In general, the Highly Diversified segment uses verbal, rather than printed, media to a greater extent than do the other segments. The printed medium on which they rank highest emphasizes one's immediate surroundings (local newspapers). Even the high-ranking content usage categories for newspapers are largely focused on local issues, namely, social news and real estate sections. They also rank relatively high in their use of financial newspapers.

Their movie viewing appears to be influenced by the fact that a high percentage of the adults in this segment are in households with children present. The mix of movie types on which they rank high ranges from horror films to documentaries and religious films.

PTV seems to play an important role, not only in satisfying the diversity of needs exhibited by various family members, but also in relating directly to this segment's own needs for intellectual stimulation and growth, and to their concern for the development of their children.

Family-Integrated Activities

High percentage of adult women with young children. Strong interest in home and in family interactive activities—household activities and management and indoor games. High need for family ties. Child presence influences adult interest patterns.

With respect to both their public and overall television viewing, the people in this segment are only slightly above the average for the entire population (5% and 6%, respectively). In general, they use less media than do persons in the other above-average PTV segments. The only two media on which their usage ranks them in the top three segments are magazines (32% above average) and Sunday supplements (40%). They rank fifth in radio usage in which they are 22% above average. The next highest media usage ratios for people in this segment are those given above for television.

Individuals in this segment rank among the top three in their viewing of three program types: children's programs, dramas, and movies. Their relative exposure to children's programs is exceedingly high, 224% of average, the third highest index in Table 5.2. In addition to these three program types, they are also more than 10% above average in their viewing of documentaries, game shows, soap operas, and specials.

Their high viewing index for children's programs is to be expected. Members of this segment are 87% female, and 75% are adults with children. On both statistics they are higher than any other segment. Furthermore, the average age of their children is nine, the youngest of any segment. The programs in this category watched most often, relative to the population, are those PTV programs that combine entertainment with educational material, *Sesame Street, Mister Rogers,* and *Electric Company.* This segment's viewing rate for each of these programs is at least three and one-half times the average for the population studied. Parents in this segment watch these programs with their children and tend, in general, to direct their children's viewing toward those program alternatives seen as contributing to their children's intellectual and social development.

Further evidence of their interest in the development of their older children and themselves is the fact that individuals in this segment rank third in their viewing of three other PTV programs, *Nova, Masterpiece Theatre,* and *Once Upon a Classic.*

In addition, these are people who tend to select programs stressing family-oriented subject matter, even when viewing programming that is not solely child-oriented. For example:

(1) *Dramas.* Two of the three dramas on which the people in this segment rank among the top two segments in viewing frequency are *Little House on the Prairie* and *Family.*
(2) *Soap Operas.* Two of the three soap operas they watch well above average are *All My Children* and *The Young and the Restless.*

(3) *Miscellaneous.* Across all other program types, there are seven other programs for which people in this segment rank among the top two segments, namely:
- *Once Upon a Classic*
- *Wonderful World of Disney*
- *Love Boat*
- *Quincy*
- *Rossetti and Ryan*
- *Soap*
- *Late Night Movies*

The first three of these programs are appealing to both adults and children and thus lend themselves to family viewing. The latter four are not available for viewing until after the bedtimes of young children.

With few exceptions, the extremely high levels of viewing behavior of the adults in this segment are associated with programs that are either aimed toward the intellectual and social development of young children, or are focused on family participation or are about young families.

Their general pattern of usage with respect to magazines, movies, books, and newspapers also reflects the presence of children and the importance of family relationships. They rank first in their readership of both women's services and home service/home magazines. They rank first on eleven of the sixteen magazines that comprise these categories and second on two. Of these high-ranking magazines, three of them deal with children, namely, *Parent's Magazine, Baby Talk,* and *Mother's Manual.* They also rank first in their attendance of children's movies. The people in this segment frequently read psychology, self-help books, and "how to" books. Finally, when it comes to newspapers they are above-average readers of personal advice, gardening, and advertising material.

The behavior of the members of this segment is quite consistent with the previous characterization of its PTV viewing behavior, interests, needs, and demographics. Their television viewing appears to be influenced more by children than that of any other segment. This finding is more than a reflection of the fact that this segment contains households with children and teenagers, as other segments share this characteristic. What makes this segment unique is the extent to which the children impact on the behavior of its adult members.

Television appears to play a very definite and sharply focused role in the households of the Family-Integrated Activities segment. Through the selection of programs that have a broad range of appeal across age groups, they are able to use television as a means of bringing parents and children together to share the viewing experience, at least during the hours when the children are awake. This is markedly different from those households in which television is used as a "babysitter" or as a vehicle for keeping children out of the way while their parents pursue separate interests and activities.

For all five of the above-average PTV viewing segments there appears to be a marked consistency of function and purpose in their overall patterns of media usage. For the heavy PTV viewers the conclusions described in Chapter 4 are reinforced by their use of other media. These patterns across media may serve as one more source of ideas for programming strategies for increasing their usage of PTV.

Conclusions

For people in the five above-average PTV viewing segments, data on other media reinforce, but do not change, the interpretations discussed in Chapter 4.

Individuals in both the Arts and Cultural Activities and Cosmopolitan Self-Enrichment segments are the personification of what many correctly perceive to be PTV's primary audience. They are characterized by:

(1) extremely high levels of education;
(2) unusually high needs for intellectual stimulation and growth;
(3) selective usage of all media to broaden their intellectual and cultural horizons; and

(4) a bias toward media that emphasize somewhat more abstract, culturally upscale material.

The types of media they use in addition to public television are quite consistent with their PTV viewing behavior. Their general life style appears to emphasize a desire to broaden their intellectual and cultural horizons. We believe that these values are sufficiently important to these people that they permeate most, if not all, of the activities in which they engage. For them, broadening their intellectual horizons is seen as a way of life, not as an activity that one does for a somewhat more narrowly defined purpose (e.g., to help educate one's children).

For those in the remaining three above-average PTV viewing segments, the use of media other than PTV leads us to conclude that PTV plays a more limited role in their lives. Each of these segments is discussed briefly below.

Family-Integrated Activities. For the people in this segment, PTV does not appear to be as important for one's personal development as for those in the aforementioned two segments. It appears to serve more as a vehicle aimed at the intellectual and social development of their young children and as a tool to foster family viewing.

Highly Diversified. These individuals seek exposure to abstract, culturally upscale material. However, this search does not prevade their overall pattern of media habits as it does for those in the Arts and Cultural Activities and the Cosmopolitan Self-Enrichment segments. We hypothesize that the people in this segment use PTV as a "school-like" substitute for personal and family development. It appears to serve more as a means of cultural and intellectual elevation rather than as an expression of an already internalized set of values.

News and Information. In contrast to the individuals in the other four segments, those in this segment do not appear to use PTV or other media out of a desire for personal development or for exposure to abstract, culturally upscale material. These people use PTV and other media to gain knowledge about a wide range of subjects in order to make themselves more socially stimulating and to be better able to converse with others.

For four of these five segments the desire for exposure to abstract, culturally and intellectually upscale material plays an important role in attracting their members to PTV. PTV is unlikely, however, to obtain substantial audience penetration among the below-average PTV viewing segments in a world of voluntary viewing if its appeal is principally restricted to material of this type. Most of the below-average viewing segments do not use PTV, or any of the mass media, for intellectual or cultural stimulation. These issues will be discussed in greater detail in Chapter 6.

Note

1. Readers interested in more detailed data for each of these media by interest segment are referred to the pages indicated in Frank and Greenberg (1980): book readership by type (p. 150); commercial and public television viewing by program (Appendix F, pp. 318-337); magazine readership by magazine (Appendix G, pp. 339-353); movie viewing by type (p. 151); newspaper readership by section (p. 152); radio listening frequency by program type (p. 153).

6

Media Usage Among Below-Average PTV User Segments

The discussion of the media behavior of people in the nine below-average PTV viewing segments starts with the segment whose members report the least exposure to PTV, the Family- and Community-Centered segment, and finishes with the segment whose members, though below average, are most exposed—the Athletic and Social Activities segment.

Due to the sparseness of their PTV usage, clues as to the types of programming that might attract individuals in these nine segments must come primarily from their usage of other media. Therefore, the discussion that follows presents a detailed analysis of their commercial television viewing as well as their usage of books, magazines, movies, newspapers, and radio.

In the process of writing this chapter we have debated what, beyond reporting and interpreting these facts, we might say regarding PTV program development to address these segments. Both authors are specialists in research. Neither of us has experience in the creative end of writing or evaluating program scripts, let alone producing and airing them. Though it is tempting to avoid any suggestions for program content or format and leave such judgments completely to those with programming experience, we have decided to include a discussion of the implications of our findings for programming. Our discussion consists predominantly of illustrations, which we hope will help those skilled in the creative aspects of program development and production to more easily relate to and internalize the material contained in this volume. Our commentary is meant to serve as a primer to facilitate the process of using this type of research as an integral part of the planning process for PTV programming.

The discussion in this chapter is based primarily on the same data as those used in Chapter 5, Tables 5.1 and 5.2.

Segment Findings

Family- and Community-Centered

> Employed blue-collar/white-collar adult males. Married, living in nonmetropolitan areas. Broad interests, including outdoor activities as well as religion. Very strong need for family ties.

Overall. Individuals in this segment score below average on usage of all media except newspapers. Their newspaper readership scores range from 7% to 15% above average for all five types of papers. The next highest scoring medium for people in this segment is overall television (87% of average). This score is depressed considerably by their exceptionally light viewing score for PTV (48% of average). With respect to the remaining media, their scores are not only below average but markedly so, that is, 80% of average or less.

Television. They score above average on only four program types, namely: news/commentary, daily news shows, sports, and other (due to their high score on religious programming). For none of the nineteen program types are these people among the top three segments in their viewing. For eleven of the program types, they rank among the lowest four segments in their viewing.

It is likely that their relatively low usage of television, as well as other media except newspapers, is due in large part to the family and community activities that compete for their time. This segment also shows evidence of a high degree of involvement with religion, which may reflect values that cause them to find much of what is aired on television or depicted in other mass media to be objectionable. Their involvement in religion is evidenced by the following:

(1) Their religious interest factor score (Chapter 3) places them among the top three segments on this interest.
(2) Their television viewing of programs with religious content is well above average.
(3) Their usage of religious material in books, movies, and radio ranks them second of fourteen segments on *each* of these three media.
(4) The television program types on which they rank the highest, sports and news, are apt to be the least offensive to their religious beliefs and values.

The religious interest of this segment is not by itself, however, a totally satisfactory explanation for their restricted viewing of television, even if it is partially responsible. Certainly, there are many program types other than news and sports that do not contain objectionable material. Also, there are several other segments who evidence equally great involvement in religion, but spend more time viewing television and/or using other media.

Other Media. This segment's members are above average in their usage of Sunday papers, Sunday supplements, daily papers, local weeklies, and financial newspapers. On all five types of newspapers they rank sixth or above. This pattern of usage is consistent with their broad range of interests and their orientation toward family and community activities. Newspapers, more than any other medium, contain material tailored to local interests, while at the same time they encompass a relatively broad range of subject matter. With respect to newspaper section readership, the members of this segment rank among the top four on local news, sports, world news, and national news.

Though their overall usage of books, movies, and radio is well below average, when people in this segment are asked specifically about religion as a content category in relation to each medium, they rank second of the fourteen segments on their exposure to it.

In spite of their broad interests, members of this segment are relatively light users of magazines. They do not rank first or second in their usage of any magazine type. They rank third on only one category (outdoor) and fourth on four others: sports, business/finance, automotive, and mechanics. These reading habits are closely associated with a subset of their interests. That is, this segment is composed of people who rank second on the Reaping Nature's Benefits and Investments factors and third on Professional Sports and Mechanical Activities factors.

Several of their high-ranking interests (e.g., Community Activities, Religion, and Crime Society) either do not correspond closely to existing types of national magazines or are published in too intellectualized a form to be acceptable to members of this segment. Local newspapers may be better suited than currently available magazines for pursuing these interests.

PTV Commentary. It is quite likely that PTV has at least one major appeal to people in this segment, namely: the presence of little or no objectionable material. By the same token, given their family and community interests, there is little on PTV that is tailored to serving their needs.

There are excellent children's programs on PTV. However, individuals in this segment have children who are, on average, three years older than those in the Family-Integrated Activities segment, who report high levels of PTV viewing.

Certainly more attention to programming with religious content might attract more viewers from this segment, as might programs emphasizing the developing and sharing of traditional values within families—especially families with adolescents or early teens. Finally, programming focused on local community activities might, and possibly already does, attract people in this segment. Our study did not measure the viewing of such local programs and, therefore, may underestimate the overall PTV usage of this segment.

Elderly Concerns

Oldest segment, high percentage of retirees, widowed, few children. Very few interests include religion and news and information. Focus on maintaining sense of social integration and belonging in absence of direct interpersonal contact. Needs to overcome loneliness and lift spirits. Low need for intellectual stimulation.

Overall. The people in this segment score above average on their usage of only two media, overall television and local newspapers. Their viewing of PTV is 50% of average (only 12.5% report watching once a week or more), and hence their high score for overall television is due to their heavy watching of commercial television and not PTV.

Their narrow range of media usage is consistent with their interests. Individuals in this segment score below average on fifteen of the eighteen interest factors. The other three are Religion, News and Information, and Household Activities and Management. These interests, together with their exceptionally high needs for Greater Self-Acceptance (e.g., to overcome loneliness and lift one's spirits) and to be Socially Stimulating, substantially influence the content of their media usage.

Television. The people in this segment have higher levels of viewing behavior than do those in any other segment for dramas, game shows, soap operas, and religious programming. In addition, they are substantially above average for the population (20% or more) in their viewing of news/commentary, daily news shows, talk shows, and variety shows. At the opposite extreme, they are extremely light viewers of children's programs (34%) and are well below average in their viewing of sports, theatrical performances, science fiction, musical performances, movies, and specials.

Their above-average viewing of news and religious programming comes as no surprise, as these content areas match their expressed interests. Both their interests and viewing of these types of programs probably help to satisfy their needs for social integration and acceptance.

Their exceptionally high levels of viewing of the other program types listed above also appear to reflect these same needs. A common denominator of these program types is that they provide vehicles for vicarious participation. Heavy viewers of soap operas are known to become involved in a very personal way with the characters in the stories and with the problems in their lives. The hosts and celebrities of game shows and talk shows provide a kind of vicarious relationship. They tend to be stable, accepting figures with highly predictable (bordering on ritualistic) patterns of behavior. As such, they appear to offer surrogate friends, or at least acquaintances, for this segment that appears otherwise lonely and spends much of its time isolated from real-life support groups.

Many of the lead personalities associated with the programs viewed by this segment can be characterized as nonthreatening in that they exhibit an interest in, and an almost uncritical acceptance of, others. For example, *The Waltons* and *The Lawrence Welk Show* feature very nonthreatening personalities to a group that is very much in need of acceptance at this point in their lives.

There are, of course, other factors, in addition to needs, that impact on their overall viewing behavior. For example, the above-average television viewing by the people in this segment is, in part, a result of the fact that they are more apt to be at home and available for viewing, especially during the day, than are those in any other segment. Their light viewing of certain of the talk and variety shows is related not only to their content, but also to their schedule. Given that people in the Elderly Concerns segment have an average age of 61, it is reasonable to assume that they are less apt to stay awake to watch late-night shows such as *Fernwood 2 Night* and *Saturday Night Live,* even if their content were attractive to them (which it is not).

In the following paragraphs, each of the program types with above-average viewing by this segment, except for news and religion, is commented upon:

(1) *Dramas.* The fact that this segment's members have the highest viewing index for drama is principally due to their exceedingly heavy viewing of two of the eight programs in this category, *Little House on the Prairie* and *The Waltons*. In contrast, they view *Rafferty* and *The Fitzpatricks* to a lesser extent than do members of any other segment. Both of the more heavily viewed programs emphasize family solidarity and acceptance.

(2) *Soap Operas.* Members of this segment watch daytime soap operas considerably more than the average for the population. This attraction to "soaps" is broad-based. There are eleven programs in this category and these people rank above average in their viewing of all, ranking first on six programs and fourth or above on the remaining programs.
(3) *Game Shows.* As in the case of soap operas, people in this segment are well above average in their viewing of all nine programs in this category. Of these, they rank first on three of them and fifth or higher on the remaining shows.
(4) *Talk Shows.* Within this category there are substantial differences in this segment's viewing behavior across programs. They rank first in their viewing of Merv Griffin, Mike Douglas, *Dialing for Dollars,* and *Women.* They are below average in their viewing of *Fernwood 2 Night,* Johnny Carson, and Phil Donahue. There appears to be a pattern of watching personalities who tend to be less critical and more accepting of others.
(5) *Variety Shows.* The one variety show that the people in this segment watch more than do those in any other segment is Lawrence Welk. Next in their above-average viewing are the Carol Burnett and Andy Williams' shows. They watch *Saturday Night Live* less often than does any other segment, and are well below average in their viewing of *The Richard Pryor Show.*

This general profile of program viewing reinforces our interpretation of their above-average viewing of *Women* and *Upstairs, Downstairs* on PTV in Chapter 4.

Other Media. The usage of other media, though relatively lower than that for television, is nonetheless quite similar in function. The need to stay in touch with familiar people, places, and things is reflected in their listening to all-news radio stations and reading the world news and advertising sections of newspapers.

Though their overall use of books, movies, and radio is well below average, they score high on all three as a source of religious content.

The three types of magazines on which they rank highest are: women's services, home service/home, and romance. On none of these types, however, are they above average in their readership. The particular women's services magazines on which they have higher ranks (seventh or eighth) are *Good Housekeeping, Family*

Circle, and *Woman's Day.* These are all magazines related to home maintenance activities, in contrast to magazines in this category such as *Ms., Parent's Magazine, Baby Talk,* and *Cosmopolitan* on which they rank twelfth or lower. This is consistent with the fact that Household Activities was one of the only three above-average interest factors for people in this segment.

The romance magazines on which they have relatively high ranks (fifth out of fourteen) are *Modern Romances* and *True Story.* Though only a small proportion of the people in this segment frequently read any of the romance magazines (7%), we would hypothesize that those who do are using them to cope with their loneliness in much the same manner they use soap operas, game shows, and talk shows on television.

PTV Commentary. Virtually none of the 22 PTV programs included in the study address this segment's members' interests in religious content, or their needs to cope with loneliness. *Women* and *Upstairs, Downstairs* are as close as PTV has come to dealing with these latter needs within the coverage of our survey. Had *Over Easy* been included it would have probably scored even better in reaching these people than the above programs.

As programs such as *Over Easy, Upstairs, Downstairs,* and *Women* illustrate, it is feasible to develop programs that address the needs of this segment and yet maintain PTV's standards for excellence. It is also clear that to reach these people more often PTV must feature individuals and social contexts (e.g., family settings) that are not only nonthreatening, but designed to facilitate a sense of personal identification and participation. Continuity of characters and themes can be an extremely important element in achieving these goals. Religious content offers another means of attracting them. As is clear from the wide variety of commercial television programs they view, there exists a wide range of program formats that might help increase their involvement with PTV. Just as the same children who watch cartoons on commercial television can be attracted to the excellent children's programs on PTV, the members of the Elderly Concerns segment, who watch soap operas and game shows, can be attracted to quality programs on PTV if their basic needs are properly recognized and met.

Money and Nature's Products

Older males with a higher proportion being rural and retired. Interests in passive activities that obtain some form of tangible return or product—fishing, hunting, investments. Low interest in active physical activities—camping out, participant sports—as well as culturally upscale or abstract—classical arts, international affairs. Somewhat complacent, but need interpersonal contact and support, especially from their families.

Overall. The people in this segment do not rank among the top three segments in their usage of any of the media studied. Their highest relative score is for reading daily newspapers (10% above average). Their second highest score is for reading Sunday papers, on which their usage is equal to the average for the entire population. Their overall viewing of television is next (98% of average), with viewing of PTV at only 57% of average.

Television. The program types with above-average viewing by the members of this segment reflect three distinct elements associated with their interests, needs, and demographic characteristics:

(1) interest in outdoor and nature-related activities;
(2) need for social contact, support, and respect; and
(3) interests in investment- and business-related subject matter.

Though it is useful to conceptualize their viewing behavior in terms of these three elements, as will be seen, programs that incorporate any one element often serve some of the same needs and interests as those in others.

(1) Outdoor Content. Their above-average viewing of adventure, documentary, and drama program types is principally accounted for by programs associated with nature-related settings. These include:

(a) *Adventures—The Life and Times of Grizzly Adams* and *Young Dan'l Boone*, as opposed to *Six Million Dollar Man*, *Nancy Drew*, or *The Hardy Boys*.

(b) *Documentaries—Last of the Wild*, as opposed to *Age of Uncertainty*.
(c) *Dramas—Oregon Trail* and *The Waltons*, as opposed to *Family* and *The Fitzpatricks*.
(d) *Sports*—Above average on all professional sports—spectators, not players.

(2) Social Contact, Support, and Respect. Associated with this set of needs is the above-average viewing by this segment of crime dramas, talk shows, and variety shows. In the programs they view relatively often, the story tends to be focused on a single personality, close in age to the members of this segment, who is socially integrated in terms of contact, support, and respect, namely:

(a) *Crime Dramas—Kojak, Hawaii Five-O, Barnaby Jones*, as opposed to *Starsky and Hutch, Rossetti and Ryan*, and *Baretta*.
(b) *Talk Shows*—Johnny Carson is viewed more often by this segment than by any other, except for the News and Information segment. Less frequently viewed programs in this category are confounded with scheduling differences as many talk shows are aired during the daytime when the large majority of this segment is unavailable for watching television.
(c) *Variety Shows*—Lawrence Welk, Carol Burnett, and Andy Williams, as opposed to Redd Fox and *The Richard Pryor Show*.

We believe that one of the common needs of individuals in this segment is for the continued support and reinforcement of the more traditional values associated with American life. Virtually all of the main characters in the programs discussed in this and the preceding section tend to personify some or all of these values.

(3) Investment/Business Content. Though they are above-average viewers of daily news shows (13% above average), they are considerably higher in their viewing of news/commentary programs (60% above average). Their viewing of these program types is related to their interests in financial matters and business, as described in Chapter 3. In addition, there is some tendency for them to be somewhat higher than average on programs that emphasize individual personalities and their evaluations. Such

programs include *Wall Street Week* (a PTV program), *Meet the Press,* and *Face the Nation.*

Given the nature of this segment's interests in business, much of their nonfictional television viewing can be explained by a desire to be knowledgeable about what is going on in the world around them. In today's economy, success in business and in investments is often related to one's ability to be tuned in to the social and political, as well as economic, trends of our society. This segment's above-average viewing of daily news, news commentary, documentary shows, talk shows, and even sports programs helps to maintain a high level of awareness of these trends.

Other Media. Their general usage pattern for other media is consistent with their usage of one or more types of television programming. For example:

(1) *Newspapers.* The only newspaper sections on which their reported readership ranks among the top three segments are real estate, business, and travel. Readership of the first two of these is probably an expression of their interest in Investments. Their relatively heavy readership of travel sections and travel books may relate to travel opportunities associated with their outdoor interests.

(2) *Magazines.* The only magazine type ranked above fifth for people in this segment is outdoor, on which they rank second with 29% readership. Of the five magazines in this category (*Argosy, Field and Stream, Guns and Ammunition, Outdoor Life,* and *Sports Afield*), the only one they read more frequently than those in other adult male segments is *Outdoor Life.* The next three highest ranking readership categories are automotive, mechanics, and business/finance.

PTV Commentary. The people in this segment appear to respect experience and success, and to value the opportunity to be exposed to people who have obtained them. This is especially true if the successful personality is a mature individual male whose achievements reflect what this segment perceives to be solid, traditional American values.

Increased programming related to business, national, and local affairs might well attract additional members of this segment into the PTV audience. This would be especially likely if the material were focused on prominent people and how they were approaching and/or coping with significant problems of interest to this group of viewers.

This same general tactic of providing in-depth coverage of an issue or an unfolding event from the perspective of a major figure with direct, immediate involvement might also be extended to the outdoor interests of these people. Outdoor-related material is more apt to be interesting to this segment if seen through the eyes of an experienced professional such as a fisherman or hunter.

Their interest in strong personalities that reinforce traditional values in American life might lead them to be interested in programming that incorporates either historical or current events of major significance in which:

(1) strong personalities are on opposing sides;
(2) the issue is focused on traditional social values and whether or not they will be maintained; or
(3) the issue is developed through the eyes of people who were directly involved.

One illustration might be a series of programs presenting major military battles that were crucial turning points in previous wars. Such a series might focus on the military and/or the political context in which the combat was occurring as seen through the eyes of the participants involved, particularly the leaders.

The same general approach could also be applied to a broader range of subject matter such as:

(1) prominent political contests—the strategies of winners and losers;
(2) major changes in, or attempts to change, the U.S. Constitution;
(3) prominent legal cases relating to business; and
(4) discussion of strategies for combating crime, criminals, and criminal organizations.

Detached

Low socioeconomic profile. Extremely few interests and activities and few psychological needs satisfied by them. Low scores on needs related to both intellectual stimulation and interpersonal contact and support.

Overall. Members of this segment score below the population average on all of the media included in our study. Within this context their highest scores are for Sunday newspapers (95% of average), radio (93%), and overall television (93%).

Television. The Detached segment is at or below average in their viewing frequency of all nineteen program types. This finding is especially interesting in the context of Table 5.1, which indicates that the Detached segment has a television exposure index that, while below the population average, is larger than that for five other segments. The other segments, however, tend to be much more selective in their television viewing and thus score high on some program types and very low on others. The Detached segment is relatively nondiscriminating in its program selection and, as a result, their below-average viewing frequency is reflected in all categories.

Their highest scoring four program types are soap operas with an index of 99% of average, movies (93%), science fiction (93%), and crime drama (97%). Their lowest are news/commentary (66%), documentaries (59%), theatrical performances (46%), and specials (39%).

In addition to being oriented to less intellectually demanding or informative program content, the people in this segment use television largely as a vehicle for escape. Though their need scores for Escape from both Boredom and Problems, as discussed in Chapter 3, were neither exceptionally high nor low in relation to the *other segments,* they, along with the need for Status Enhancement, were among the top three needs for *people in this segment.*

Other Media. The only content element associated with media other than television that offers further insight into the people in this segment is their attraction to Black magazines and radio programming, along with their viewing of PTV's *Black Perspectives on the News.* This is not surprising, given that this segment has the third highest percentage of Black members. However, though their use of Black magazines and radio programming is relatively high compared to their own use of other media content categories, they do not rank among the top four segments.

PTV Commentary. This segment would be extremely difficult to attract to PTV. Not only is their overall use of all media relatively low, but they show little evidence of selectivity in that usage, except for a slight attraction toward content featuring Blacks.

Competitive Sports and Science/Engineering

Teenage male students with interests in male-associated mechanical activities and competitive athletics. Avoidance of female-oriented subjects and interests. High on needs for unique/creative accomplishment, intellectual stimulation and growth, status enhancement, and escape from boredom. Low needs for understanding others and for greater self-acceptance.

Overall. Members of this segment score well above average on their usage of movies (72% above) and, to a much lesser degree, ranging from 1% to 8% above average, on their usage of Sunday papers, magazines, books, and overall television. Their PTV usage score is only 67% of average.

Television. The overall television usage score understates the attraction that certain types of television content have for these predominantly adolescent males. The most distinguishing characteristic of their viewing behavior is that their viewing of sports programming is 105% above average. This is extreme, not only in relation to their own viewing of other program types, but also in comparison to that for other segments. Their index for sports viewing is the sixth largest index reported in all of Table 5.2. They rank first on all seven of the sports programs included in the category.

In addition to sports, they have viewing indices that exceed 115% on three other program types: children's programs, situation comedies, and science fiction. In the case of children's programs, the people in this segment are especially heavy viewers of cartoons. They are above average on all three programs that comprise the science fiction category, ranking somewhat higher on *Logan's Run* and *The Man from Atlantis* than on *The New Adventures of Wonder Woman*.

Of the thirty programs in the situation comedy category, they rank first or second in their viewing of the following ten:

- All In the Family
- Barney Miller
- Busting Loose
- Carter Country
- Good Times
- Happy Days
- M*A*S*H
- San Pedro Beach Bums
- Welcome Back, Kotter
- What's Happening

The five lowest-ranking situation comedies for the people in this segment are:

- Sanford and Son
- The Betty White Show
- Maude
- Rhoda
- The Tony Randall Show
- We've Got Each Other

All ten of their heavily viewed situation comedies incorporate one or more of the following characteristics:

(1) the absence of strong male or female authority figures (e.g., *Good Times, Happy Days*);
(2) mocking of male authority figures (e.g., *All in the Family, Barney Miller*); and/or
(3) strong male teenage heroes (e.g., *Happy Days, Good Times, Welcome Back, Kotter, San Pedro Beach Bums, What's Happening*).

The low-ranking situation comedies have, with one possible exception, one of two characteristics:

(1) female personalities in strong assertive roles (e.g., *Betty White, Maude, Rhoda*); and
(2) male figures in weak roles (*We've Got Each Other*).

In general, the types of situation comedies that are most appealing to this segment bear in some way upon the adolescent male's conflicts with his struggle for independence from parental and other adult authority figures. The heavily viewed programs in this category tend to provide a light and humorous treatment of this conflict and, in doing so, may provide considerable tension release to those whose lives are constantly affected by it.

The people in this segment are among the lowest three segments in their viewing of the following program types:
- Talk shows—76% of average
- Dramas—72%
- News/commentaries—48%
- Others (religion)—39%
- Soap operas—29%
- Theatrical performances—19%

The general characteristics indicated by their below-average viewing of these program types are:

(1) an avoidance of intellectual, cultural, informational or abstract material, especially reflected in their light viewing of news/commentaries, talk shows, theatrical performances, and religious program types; and
(2) possible discomfort with family shows that often depict traditional authority relationships within the family and cooperative coping with life's more serious problems. Such presentations of family life may be threatening to a group that often is in the midst of family conflict and having great difficulty in resolving it. This interpretation is supported by the fact that within the general drama category, the people in this segment are among the lowest three segments in their viewing of *Little House on the Prairie, Big Hawaii,* and *The Waltons,* while they are among the top three segments in their viewing of *Lou Grant* and *The Fitzpatricks.*

Individuals in this segment appear to be attracted to male-dominated program content in which the male figures are admired and looked up to because of their physical and/or social prowess, but not for their role as parents. They tend to avoid programming involving traditional authority relationships (unless treated as comedy), whether they are male- or female-related. For them, television appears to be an entertainment or recreational medium with little use dedicated to seeking information or broadening their intellectual or cultural horizons.

Other Media. The people in this segment favor book and movie content types that appear to provide a means of escape and fantasizing, as does much of their television viewing. They have high usage rates of humor and science fiction books as well as similar categories of movies. In addition, they are also high on such movie types as crime and spy, horror, disaster, and westerns. Their relatively high readership of newspaper comics may well satisfy the same needs.

Their usage of most newspaper content sections is quite low, ranking thirteenth or fourteenth for nine out of fifteen sections. However, on one other section besides comics they rank first, namely, sports. Sports also receives a rank of one with respect to their radio usage. This is to be expected given their sports-oriented interest pattern.

Their magazine readership, in addition to emphasizing sports, is high when it comes to automotive and mechanics types of magazines.

Members of this segment frequently use the radio to listen to disco, top hits of the week, and rock music. Finally, they also rank high on a broad range of radio listening including ethnic programming, drama, and talk shows.

PTV Commentary. One could make an argument that the least recognized and catered to minority audience in PTV is not children, Blacks or even Hispanics, but adolescents. These people, who are no longer children but have not yet developed the interest and need patterns that characterize the adult segments, seem to "fall between the cracks" when it comes to PTV programming strategy.

What little use people in this segment currently make of PTV is, in our judgment, primarily related to either school assignments (*Age of Uncertainty*) or to coincidental viewing in the presence of younger children (*Mister Rogers, Electric Company*). Otherwise, they generally avoid the kind of intellectual, cultural, informational, or abstract material that constitutes much of what is aired on PTV.

One possible means of increasing PTV usage by this segment and by other Youth Concentration segment members still in school is the development and integration of programming related to high school curricula. The creation of visually dramatic educational material related to standard science, history, or literature courses could stimulate viewing by this segment, particularly if strongly encouraged or required by the schools. PTV might even consider supporting the development of classroom materials, such as texts, workbooks, examinations, or other presentation materials that would complement the use of one or more PTV programs in a classroom or home setting.

Clearly, there is also a potential for attracting members of this segment to sports programming, as their appetite for the subject appears to be almost insatiable. Such programming need not be restricted to the broadcasting of live events, but could incorporate presentations of the history of sports, biographical material on famous athletes, discussions of strategies for winning, and other related topics.

Still another approach might be to attempt to develop programming that combines education and entertainment in a manner consistent with the everyday life of people in this segment (and those in the other Youth Concentration segments, especially the Athletic and Social Activities segment). This could be the equivalent of a *Sesame Street* aimed at adolescents, rather than young children. Substantial increases in PTV exposure among this and other Youth Concentration segments are not likely to be achieved unless PTV seeks innovative ways of attracting them. Understanding their patterns of interests and needs, however, provides a basis for trying to reach them, should PTV wish to do so.

Indoor Games and Social Activities

Young, low-income females. Interests in activities, especially indoor games. Low interests in most subject matter areas. Nonintellectual. High needs for status enhancement and the need to be socially stimulating.

Overall. The young women in this segment rank second out of the fourteen segments in their readership of books (62% above average). They also score above average in their usage of movies (31%), commercial television (14%), and magazines (8%). They are below average in their usage of radio and all types of newspapers as well as PTV.

Television. Not only do members of this segment watch a substantial amount of television, they view considerably more television than do members of the Athletic and Social Activities segment, which is also composed primarily of young women.

These individuals are above-average viewers of ten of the nineteen program types. On only six program types are they more than 20% below the average for the population in their viewing behavior. The six types are:

- Talk shows—76%
- Others (religion)—64%
- Theatrical performances—60%
- News shows—daily—46%
- Documentaries—44%
- News/commentaries—27%

Like the Competitive Sports and Science/Engineering segment, they clearly avoid program content that is information-oriented or culturally enriching. While their viewing of musical performances is quite high, this is attributable to their ranking first in the frequency of watching *American Bandstand* and second on *Soul Train*. Otherwise, their general viewing pattern is less selective than that of almost any other segment. They watch an above average amount of virtually every other type of programming aired on television.

The average person in this segment is a female adult in a major transitional stage of her life. At age 22, she has, in most cases, moved from her parents' home, completed her schooling, and is in the early stages of establishing her own home. The diversity of viewing behavior for this segment undoubtedly reflects a mixture of young women, some of whom are teenagers and some of whom are married with young children and many hours available at home.

In particular, they are well above average in their viewing of:

(1) youth-oriented program types, such as science fiction, musical performances (due to their heavy viewing of *Soul Train* and

American Bandstand), adventures, crime dramas (highest ranking are *Charlie's Angels, Starsky and Hutch, Switch*), and children's programs;
(2) daytime programs, such as game shows and soap operas; and
(3) situation comedies, including an unusually broad variety of programs.

This segment has by far the highest viewing index for situation comedies of all the segments, and they rank high in their viewing of virtually all of them.

It is also notable that 21% of this segment's members are Black. This percentage is the second highest among the fourteen segments and undoubtedly contributes in large part to its high rating for musical performances (*Soul Train*) and for its above average viewing of variety shows, which include *The Redd Fox* and *The Richard Pryor Shows*.

It is clear that television plays a significant role in the lives of the Indoor Games and Social Activities segment. Its members appear to be at a stage in their lives when their away-from-home social activities are restricted and when, at least for the time being, television viewing constitutes one of their major sources of entertainment.

Other Media. This segment's members report the second highest number of books read per year, a figure 62% above average, and they are 8% above average in their use of magazines. This pattern does not extend to other print media, however. Their readership of all forms of newspapers—daily, weekly, and Sunday—is below average. The only section of newspapers on which they rank above the average is the comics section. They are well above average in movie attendance (31% above) and are the lowest of the three youth concentration segments in their use of radio (21% below average).

An examination of the detailed media usage of the Indoor Games and Social Activities segment reflects a pattern consistent with this group's mix of adolescent girls and young women in the early stages of adulthood. The types of magazines that they read at well above average levels include romance, women's services, home service/home, Black, and especially *TV Guide* on which they rank above all of the other thirteen segments. They are especially low in their readership of business/finance, news, and outdoor magazines. Their book usage reflects relatively heavy readership of fiction categories, especially mysteries, and light readership of all categories of nonfiction books. Their movie attendance is especially high for comedies and love and romance films and their radio listening is well above average for disco, rock, and top hits, while they are ranked below all other segments in listening to news stations.

This segment exhibits a pattern of media usage that is somewhat different from the other two Youth Concentration segments. The Indoor Games and Social Activities segment watches more television and reads many more books, while attending fewer movies and listening to less radio. They use the media largely for entertainment, except for the magazines they use to learn about how to become a homemaker.

While they rank below the population average in their use of the media for intellectual and cultural enrichment, they score well above the other Youth Concentration segments on such behavior, while exhibiting less interest in such escapist material as horror, disaster, and science fiction films.

PTV Commentary. As indicated above, the members of this segment span a range of late-adolescent girls and young adult women. Their media behavior reflects an effort to satisfy the interests and needs common to both groups and to help them make the transition into adulthood.

The Indoor Games and Social Activities segment could probably be attracted into the PTV audience by programming designed to help young women who are learning the role of homemaker to adjust to that role. Programs could provide information and instruction in an entertaining manner on cooking, gardening, housekeeping, and related subjects. Note, however, that in order to appeal to this segment, such programs would have to acknowledge their youth and inexperience. While these young women may want to learn to cook, they are probably not ready for gourmet training.

They are also available during the day for programming that provides an alternative to soap operas and game or quiz shows. During this period, they might respond to programs that provide alternative forms of entertainment that incorporate either contemporary music or comedy, since these elements appear to be present in much of their current television viewing. The development of such programs on PTV, in order to be successful in attracting this segment, should feature young adults, especially women, and should deal with topics and issues that are of special concern to people of this age group, without appearing to be pedantic.

This segment ranks fourth in the amount of television they watch, but only ninth in their viewing of PTV. Consequently, their potential for increased viewing of PTV is significant should they be identified as a target audience segment and efforts be made to reach them. Their relative youth adds to their attractiveness, as once they become familiar with and form the habit of viewing PTV, they are likely to remain in the audience for years to come, even as their interests, needs, and life styles change and they shift into other audience segments.

Mechanics and Outdoor Life

Young adult, blue-collar males whose interests focus on noncompetitive activities emphasizing personal, physical accomplishment—e.g., auto repair, fishing, camping—interests that, by their very nature, do not require emphasis on interpersonal cooperation or support. High on needs for escape and unique/creative accomplishment.

Overall. The people in this segment make exceptionally heavy use of movies and radio, whereas their usage of all other types of media is below average. When one examines the content of their overall media behavior, one sees striking similarities in the types of material to which they are attracted as well as in the types they avoid. These patterns are present despite large differences in their relative usage of different media.

Television. They are above-average viewers of only four program types, namely: science fiction (27% above average), adventures (24%), crime dramas (9%), and movies (8%). The first three of these all provide vehicles for escape and fantasizing. It is also likely that their television movie selection is consistent with this pattern. We do not know what kinds of television movies they watch as they were not asked to describe them. However, their viewing of movies away from home is oriented toward escapist material of the sort contained in the preceding three television program types.

This interpretation of their viewing pattern is further reinforced by their choices of programs within each of the three types as described below:

(1) *Science Fiction.* Though they are above-average viewers of all three programs in this category, they are especially high on *Logan's Run,* next on *The New Adventures of Wonder Woman,* and least above average on *The Man from Atlantis.* It appears that the closer the program is to reality, the less they deviate from the average viewing frequency.

(2) *Adventures.* The two programs that the people in this segment view most, relative to the population average, are *Six Million Dollar Man* and *The Bionic Woman.* They report their lowest relative frequency for *Young Dan'l Boone.*

(3) *Crime Dramas.* The higher ranking programs for these people are *Chips, Charlie's Angels, Baretta,* and *Rockford Files,* as opposed to *Switch, Barnaby Jones, Police Woman,* and *Quincy.*

In general, the Mechanics and Outdoor Life segment concentrates its relatively heavy viewing on programs that involve the elements of adventure, drama, and suspense in vehicles in which the forces of good and evil are clearly drawn. These shows are relatively fast-paced and focus upon hero figures who inevitably triumph in the end. One would hypothesize that this segment would have been heavy viewers of westerns during the 1950s and 1960s when such programs were abundantly available.

There are, of course, many forms of escape, depending upon an individual's personality and needs. For some people, situation comedies provide an outlet for escape. For others, soap operas or game shows may perform the same function. One can learn as much about a segment by noting the types of programs they elect *not* to view as well as those they choose to watch.

Members of this segment score well below average on program types that contain more intellectual content or abstract subject matter. They rank among the lowest three segments in their viewing of:

- Other (religion)—13% of average
- *Soap operas—25%
- Theatrical performances—29%
- Talk shows—29%
- Musical performances—52%
- *Game shows—60%
- News shows—Daily—70%
- Sports—72%

The two program types labeled with asterisks contain shows that are aired at times when the adult, male, employed members who constitute a clear majority of this segment are typically not at home. However, the low scores on the other types do not appear to be explainable by schedule considerations. All of the remaining types contain programs predominantly shown during nonworking hours. In the case of talk shows, available during both daytime and late evening, this segment exhibits below-average viewing during both time periods. All of the remaining program types listed, with the exception of sports, contain intellectual or cultural content.

Even their low viewing rate for sports is consistent with their expressed interests and needs. It was shown in Chapter 3 that people in this segment tend to have low levels of interest in competitive activities. All of the programs contained in the sports category are competitive in nature.

Other Media. This segment's members are relatively heavy users of movies and radio (49% and 34% above average, respectively). These are the only two media on which their usage ranks them among the top three of the fourteen segments. For books, magazines, and all types of newspapers they rank ninth or lower.

For people in this segment, virtually all of the high-ranking media content areas are consistent with their interests in mechanical activities, outdoor activities, or their needs to escape from boredom and problems. They are the second highest segment on readership of "how to" books with 40% reporting themselves as regular readers, compared to 30% for the population. They rank first in their readership of the men's, outdoor, and miscellaneous (*Popular Photography* and *National Lampoon*) magazine types. They rank fourth in listening to country music and farm programs, which is probably, in part, a function of their interest in outdoor activities and their rural locations.

It is hypothesized that the need to escape from both problems and boredom for the people in this segment is an important determinant of the types of movies they attend and the books they read. For movies, they rank fourth or above in attendance of westerns, science fiction or supernatural, horror, and crime and spy films. Other than for "how to" books, their high-ranking book content categories are science fiction and humor.

It seems likely that their heavy usage of newspaper comics serves a similar escape function. The only segments that rank higher on reading the comics are two of the Youth Concentration segments.

Members of this segment are quite low in their exposure to religious content and, more generally, in usage of material with intellectual or cultural content, such as biographical books, select magazines, movie musicals, operas, dances, and educational/instructional radio programs.

PTV Commentary. Much of the media use of individuals in this segment appears to be associated with their needs to escape or to fantasize. While PTV cannot change this, it can respond, if it wishes to attract them into its audience. This could be achieved without deviating from PTV's standards of excellence by offering quality programming consistent with their interests and needs. For example, there are many excellent literary works in the areas of science fiction and crime drama. Properly scripted, and perhaps more importantly, properly advertised (e.g., emphasizing their *excitement* rather than their literary merit or historical interest), such programming could be successful in reaching these people. Other elements of program content that might have appeal to the people in this segment are those associated with away-from-home activities that emphasize personal development or physical activity such as camping out, hiking, jogging or biking.

Home- and Community-Centered

Adult females with a relatively high percentage of married homemakers. Home and local community interests. Highest needs for family ties and understanding others. Lowest needs for intellectual stimulation and for unique/creative accomplishment.

Overall. The members of this segment are above average only in their usage of three types of newspapers: local weekly, Sunday, and daily papers. They are at or below average on all other media. On no media are they below 83% of average. Hence, though below average, they are moderate, not light users of these media.

Television. They are well above average, ranking second out of fourteen, in their viewing of only two television program types, namely, soap operas (80% above average) and the "others" program category (64% above average), which is heavy in religious programming. They are above average in their viewing of seven other program types. On five of these they are within 14% of average and in two, game shows and talk shows, they are 21% and 19% above average, respectively. There appears to be no clear systematic pattern of viewing of these seven program types or of the programs within them.

However, with respect to soap operas and religious programming, there is an element of systematic behavior. Fully 35% of the members of this segment are homemakers, almost double the population average of 19%. They tend to be home during the day and to be taking care of children. In addition, these people have relatively narrow interests and needs focused on their families and communities.

In many ways, their viewing pattern is similar to that for the Elderly Concerns segment. A major difference in daytime viewing is, of course, reflected in the above-average index for children's programs among the Home- and Community-Centered segment, while the Elderly Concerns segment has the lowest index of any group on this program type. It is likely that the present segment, relatively cut off from adult companionship during the day, finds that the soap operas, game shows, religious programs, and talk shows satisfy a need for social integration in much the same way as they do for the Elderly Concerns segment.

The Home- and Community-Centered segment appears to use television primarily as described above during the day and for entertainment during the evening hours, with little effort to use it as a medium for gaining information or broadening their cultural exposure. Their viewing pattern is almost diametrically opposed to that of the Arts and Cultural Activities segment, for example, if one compares the relative high and low scores for these two segments in Table 5.2.

Other Media. The only media that the members of this segment use more than the average for the population are local weekly papers, Sunday papers, and daily papers. The newspaper sections on which they rank among the top three segments are those that provide information about subjects or activities related to one's home or local community, such as cooking, social news, personal advice, gardening, advertising, and entertainment.

This is in contrast to their newspaper readership of material related to subjects or activities outside of the home and community, where they rank low. For example, their readership of world news ranks eleventh. Only 42% of the people in this segment read this section frequently versus 56% for the population. Their national news readership also ranks relatively low (eighth), with a slightly lower-than-average percentage of the segment's membership frequently reading it.

Their high ranking of religious content for books, movies, and radio programs is to be expected, as the people in this segment rank fourth on the Religion interest factor.

When it comes to magazines, these people rank second only to those in the Family-Integrated Activities segment in their readership of women's services and home services/home magazines. Within the women's services category their highest-ranking magazines are *Family Circle, Woman's Day,* and *Ladies' Home Journal,* while lowest ranking are *Ms., Mother's Manual, Cosmopolitan,* and *Parent's Magazine.* This aspect of their readership profile is consistent with their interest in Household Activities and Management.

They have interests that are less intellectual than those of the Arts and Cultural Activities or the Family-Integrated Activities segment. Their usage of magazine content is more narrowly focused, and they have lower readership on ten of the fifteen magazine types than do those in either of these two segments. The narrower focus of their magazine usage may also be due, in part, to their interest in Community Activities, together with the fact that none of the magazines included in this study focuses on such activities. This hypothesis is supported by the fact that this segment makes more use of local weekly newspapers than do the other two.

PTV Commentary. This segment's needs for social integration during the day are similar to those of the Elderly Concerns segment. During this period, the Home- and Community-Centered segment can be attracted by programming similar to that recommended for the Elderly Concerns segment—religious programs and programs that facilitate a sense of personal identification and participation in an adult world.

This segment can also be reached through programming that focuses on local community personalities, activities, and events. In this area, local PBS stations have the opportunity to augment the coverage of local newspapers, providing greater depth and a sense of vicarious participation. It is likely, for example, that a talk show with local guests, a cooking program featuring recipes

BELOW-AVERAGE USER SEGMENTS

with a regional slant, or a gardening program talking about local flora would attract these people.

Athletic and Social Activities

Teenage females from high-income families. The youngest of all the segments. Interests in active, away-from-home, face-to-face activities. High need to escape from problems and to be socially stimulating. Low need for family ties.

Overall. The adolescent females in this segment are above average in their usage of only two media, movies (87% above average) and radio (61%). The next highest scoring media are magazines (98% of average) and PTV (97% of average). They are well below average in their readership of all types of newspapers (70% of average or less).

Television. Though they are only slightly below average on their PTV score, they are fully 20% below average when it comes to their score on overall television watching.

They are above-average viewers on only four of the nineteen program types as follows:

- Science fiction—22% above average
- Children's programs—21%
- Situation comedies—15%
- Movies—8%

At the opposite extreme, their viewing index is below 50% of the average for the population in the following categories:

- Documentaries—3% of average
- Theatrical performances—19%
- News/commentaries—35%
- News shows—daily—40%
- Talk shows—42%

Like the other Youth Concentration segments, these people are well below average in their viewing of information-oriented or cultural subject matter.

We believe that the need to escape from problems at home, combined with a low need for family ties on the part of members of this segment, is met by immersing themselves in away-from-home interests and activities. In addition, part of their above-average viewing behavior also serves an escape function, namely:

(1) *Children's Programs.* They rank among the top four segments in their viewing of *Mr. Rogers, Captain Kangaroo,* and *Electric Company.*

(2) *Science Fiction.* They are above average in their viewing of all three programs in this type though they rank somewhat higher on *The New Adventures of Wonder Woman* than on either *The Man from Atlantis* or *Logan's Run.*

(3) *Movies.* Television movie content was not specifically covered in the questionnaire; however, we do know the viewing habits of these people when it comes to attending movies. They are also above-average movie theater attenders, especially for such escape types as science fiction, horror, and disaster films.

(4) *Situation Comedies.* Their situation comedy viewing appears to serve related, but somewhat different, functions. They are among the two top segments in their viewing of:

- Laverne and Shirley
- Love Boat
- One Day at a Time
- Operation Petticoat
- San Pedro Beach Bums
- Soap
- Three's Company

Common to all of these programs are light-hearted treatments of male/female relationships in nontraditional family (or household) relationships. As such, they may provide the same sort of tension release for the members of this youth segment as offered to the Competitive Sports and Science/Engineering by their preferred situation comedies as discussed above. Consistent with this interpretation is the fact that the Athletic and Social Activities segment ranks first in its viewing of *Soap,* the ultimate satire on family life.

For this, the youngest of all the segments (average age of 18 years), television is clearly used as an entertainment medium and not as a means of keeping informed or broadening exposure to the arts. The difference in preferences among the situation comedies selected most often by this predominantly female youth segment and those selected by their male counterparts is interesting. This difference appears to reflect a subtle difference in the nature of the adolescent conflict for boys and girls. The boys are struggling to establish their independence from traditional home and community authority figures and to exert their own authority. Their situation comedy viewing is concentrated on programs that poke fun at traditional authority figures and roles. The girls, on the other hand, while confronted with a similar struggle, seem more oriented toward simply escaping from the problems of home life. Their viewing of situation comedies is focused on programs that treat family life in general in a humorous vein.

Other Media. These people read or view almost all of the same types of escapist science fiction, horror, disaster books, and movie types as do the adolescent males in the Competitive Sports and Science/Engineering segments. In addition, 58% are frequent readers of mysteries compared to 44% of the population, ranking them second of the fourteen segments. Unlike their male counterparts, however, they rank first in attending love and romance films and second on children's movies.

Like the boys, they read few newspaper sections frequently. Comics is their only high ranking category. Their heavy usage of radio is principally oriented toward music. They rank higher than any other segment in listening to disco music, popular music, rock music, and top hits of the week. They rank third in listening to rhythm and blues. The only other types of radio programs on which they rank as high as third are Black and Spanish programs.

This segment shares with the Indoor Games and Social Activities segment a high readership of women's services, romance, and fashion magazines and low readership of business/finance, news, and mechanics' magazines.

Their lack of home orientation is reflected by the fact that they score even lower than do those in the Indoor Games and Social Activities segment in their readership of home services and home-related magazines. In addition, this segment has a somewhat higher incidence of readership of outdoor and automotive magazines, which also is probably accounted for in large part by their greater away-from-home orientation.

PTV Commentary. Two of the same programming strategies suggested for persons in the Competitive Sports and Science/Engineering segment also are appropriate for persons in this segment, namely:

(1) the development of programs as well as complementary teaching aids that facilitate integrating more PTV offerings into their school curriculum; and
(2) efforts aimed at developing programs and/or program series that combine education and entertainment at a level appropriate for persons in these segments much as *Sesame Street* does for children.

Conclusions

All of the nine segments analyzed in this chapter share in common the fact that their reported usage of PTV, once a week or less, was below the average for the entire population. Though they share the same "below-average" PTV usage label, their leisure interests, needs, and media habits for other than PTV are quite heterogeneous. In fact, they are so different from one another that for PTV to appeal to each of their own special interests and needs, it would take a variety of program offerings spanning a fairly diverse range of content and styles of execution to attract substantial numbers of viewers across all nine of these segments. Their low exposure to PTV exists as a result not of one, but of a multiplicity, of interests and needs that currently are not well served by PTV.

Though our discussion has proceeded on a segment-by-segment basis, this does not imply that the programming strategies aimed at attracting increased PTV viewing in each segment are *all* of necessity mutually exclusive of each other. While there may not be a single common denominator that is apt to be equally effective across all nine segments, there are several common patterns that might serve as the basis for programming that cuts across several segments. For example:

(1) There are elements of escape and fantasizing associated with the media usage of people in the Money and Nature's Products, Mechanics and Outdoor Life, and Competitive Sports and Science/Engineering segments.
(2) Strong interest in religious content in the media is observed on the parts of those in the Elderly Concerns, Family- and Community-Centered and Home- and Community-Centered segments.
(3) Needs for coping with loneliness appear to be satisfied via involvement in vicarious relationships with media figures for many people in the Elderly Concerns as well as in the Home- and Community-Centered segments.

We do not mean to imply that PTV should necessarily respond to all these interests and needs, nor are we, in our role as scientists, in a position to impose our value system as to which interests and needs or which audience segments should receive the highest priority. What we hope to do is to help identify both the types of people and the interests and needs that at present do not appear to be well served by PTV.

The selection of target audience segments is a strategic decision that needs to be made by the Corporation for Public Broadcasting and by the Public Broadcasting System. That decision must be made in the context of the stated missions of these organizations, the political environments in which they operate, the influences of their funding sources, and the competitive arena for viewers.

We firmly believe, however, that it is crucial for the survival of PTV that a coherent strategy incorporating a definition of its target audiences be articulated. Furthermore, we believe that the audience segmentation scheme proposed in this and our previous book provides a conceptual framework for the development of such a strategy and for the subsequent development of programming specifically designed to attract viewers from targeted audience segments.

In the concluding chapter of this book, we will discuss in more general terms the role that our findings can play in developing and refining PTV programming strategy. For now, we have defined in detail the multiplicity of interests, needs, and media behavior profiles that are associated with below-average PTV exposure, and illustrated some of the directions in which PTV programming might change if one wished to increase each segment's PTV usage.

7

Public Television Program Preferences

Chapter 4 described the PTV viewing of each of our fourteen interest segments and indicated that there are two segments that are by far the most frequent viewers, three others that are above average, and nine segments that exhibit below-average frequencies of viewing. In Chapters 5 and 6, data on other media were explored on a segment-by-segment basis in order to gain insights into how PTV viewing relates to the usage of commercial television, books, magazines, movies, newspapers, and radio. The results reported in these three chapters confirmed that our segmentation structure, based on patterns of expressed interests, was useful in generating and explaining differences in media behavior across the fourteen interest segments.

While people's *behavior* with respect to other media offers some suggestions for program development to attract PTV viewing, an alternative source of inspiration lies in examining their *expressed preferences* for PTV programming. This chapter presents and discusses two nonbehavioral measures of PTV programming interests. The first measure consists of self-reported interest in seeing the offerings of various types of programs on PTV expanded. The second comprises the reactions to a series of specific new program concepts presented to each of our survey respondents for their evaluation. The data obtained on these measures are reported for each of the fourteen interest segments, along with their implications for PTV programming strategies.

Interest in Program Types and Reactions to Concepts

Ratings of Program Types

Interest in program types for broadcast on PTV was assessed in the following manner. Respondents were informed that:

> there are many different views about what types of programs Public TV should offer. The list below gives several alternatives. How interested would you be in having Public Television offer more of each of the types of programs listed below?

For each program type they were asked to express their degree of interest on a scale from 1 to 4, where a rating of "1" indicated "not at all interested" and a rating of "4" indicated "very interested." The following nine program types were presented for evaluation:

- cultural programs (e.g., drama, opera, art)
- programs similar to those on commercial TV
- music-only programs
- programs appealing to certain kinds of people (e.g., women, Blacks, Spanish-speaking, older people)
- programs about local events and issues
- news programs
- programs about special interests (e.g., home gardening, tennis, the stock market)
- educational programs
- children's program

The mean ratings for each of these program types by interest segment are reported in Table 7.1

The most favored program types for PTV are educational, news, special interests, and local events and issues. In general, the range of expressed interest for the total population across types seems relatively narrow, and the level of interest is not high. Not even the highest-rated program type, educational programs, achieved an average rating of "somewhat interested." The variation in ratings across interest segments is substantial, however, and will be discussed below.

Table 7.1 Interest in PTV Program Types[a]

Type of Program	Entire Population (average)	Arts and Cultural Activities (AF)[b]	Cosmopolitan Self-Enrichment (M)	News and Information (M)	Highly Diversified (M)	Family-Integrated Activities (AF)	Athletic and Social Activities (Y)	Home- and Community-Centered (AF)	Mechanics and Outdoor Life (AM)	Indoor Games and Social Activities (Y)	Competitive Sports and Science/Engineering (Y)	Detached (M)	Money and Nature's Products (AM)	Elderly Concerns (AF)	Family- and Community-Centered (AM)
Cultural programs	2.0	3.2	3.0	1.9	2.2	2.0	1.6	1.8	1.5	1.8	1.6	1.6	1.6	1.9	1.8
Programs similar to those on commercial TV	2.2	2.0	1.8	2.2	2.4	2.2	2.4	2.1	2.3	2.1	2.3	1.8	2.2	2.2	2.1
Music only programs	2.1	2.5	2.5	2.1	2.2	2.0	1.7	2.1	1.8	2.0	1.9	1.7	1.8	2.2	2.0
Programs appealing to certain kinds of people	1.9	2.0	2.3	1.8	2.3	1.9	2.0	1.8	1.6	1.8	1.6	1.6	1.6	1.9	1.7
Programs about local events and issues	2.4	2.8	2.6	2.8	2.7	2.4	2.1	2.4	2.1	2.0	2.1	2.0	2.5	2.7	2.5
News programs	2.5	2.9	2.5	2.8	2.8	2.4	1.8	2.5	2.2	2.1	2.2	2.0	2.8	2.9	2.6
Programs about special interest	2.4	2.7	2.8	2.6	2.7	2.7	2.4	2.4	2.1	2.4	2.2	1.9	2.4	2.3	2.4
Educational programs	2.6	3.1	3.2	2.9	2.9	3.0	2.3	2.6	2.3	2.5	2.2	2.0	2.2	2.3	2.8
Children's programs	2.3	2.6	2.7	2.5	2.8	2.8	2.2	2.3	2.0	2.5	1.8	1.9	1.8	1.9	2.4

a. Mean scale values: 4 = very interested, 3 = somewhat interested, 2 = not very interested, 1 = not at all interested.
b. Letters associated with each segment indicate which concentration it is in, namely: AF = Adult Female, AM = Adult Male, Y = Youth, and M = Mixed.

Reactions to New Program Concepts[1]

In one portion of the interview, respondents were presented with printed descriptions of eight new program concepts and were asked to rate their interest in viewing each of them, using a 4-point scale ranging from "not at all interested" (1) to "extremely interested" (4). The concepts are shown in Figure 7.1. While the concepts were not explicitly associated with PTV for our respondents, they had, in fact, been developed for consideration by PBS for broadcast. The eight concepts for evaluation were selected to represent a diversity of appeals in the hope that they would elicit differential reactions among the interest segments. In particular, we hoped to discover that people would favor program concepts consistent with their patterns of interests and activities. Not only would this reinforce the validity of the segmentation approach, but it might offer promise for the use of new program concepts as an early screening device for evaluating the potential of new programs to expand the base of PTV viewers. At the time the survey was being conducted, none of the eight program concepts had been produced and broadcast.

Table 7.2 presents the reactions to each concept by the total population and by each of the fourteen interest segments. The top number in each cell is the mean interest rating and the bottom number is the percentage of people who indicated that they were either "extremely interested" or "quite interested."

Across the entire population, the mean ratings for the new program concepts ranged from a low of 2.1 (*Hollywood Television Theatre*) to a high of 2.5 (*Your Retirement Dollar*), with the corresponding proportions of people expressing interest ranging from 33% to 49%. Again, however, the evaluations differ substantially across interest segments for several of the concepts, and it is these differences that form the basis for the discussion in the remainder of this chapter.

Reactions of Above-Average PTV User Segments

This section discusses the reactions to increased offerings on PTV of various program types and to the specific program

(text continues on p. 162)

Figure 7.1
New Program Concepts for Public Television

Just Plain Country

"Music City, U.S.A.," "City of ten thousand pickers," or "Music capital of the world." Call it what you will, Nashville, Tennessee, is the home of the Grand Ole Opry and a wealth of talent which has supplied the world with country music. This program takes advantage of this abundance of Nashville talent to appeal to television's blue-collar viewers whose language is spoken in every country song, the fans who fill the auditoriums around the country when the bus carrying their favorite star pulls into town.

Mother's Little Network

Mother's Little Network (MLN) offers something unique—a comedy born, bred, and rooted in America. Posing as an up-and-coming, family-owned broadcasting company, MLN hits the air every week with its own brand of video humor—a series of fast-paced sketches, animations, parodies, and personalities, with a format owing nothing to anyone or anything, including the meaning of its title.

MLN restores a freshness and regularity to your TV viewing. All new punchlines! All new accents! All new breaches of regional and national standards of good taste!

Sportlight

Sportlight gives viewers—in a regular, weekly, 90-minute format of live or edited coverage—the chance to see and follow a variety of first-rate competitive amateur athletic events not to be found elsewhere on American television.

On-air hosts are top journalists or well-known participants in the field, men and women who speak colorfully and incisively on the event and the issues that surround it, e.g., Bud Collins, Arthur Ashe, Donald Dell, Kem Prince, Judy Dixon, and others, who go into the background of each sport and give instructional tips where appropriate.

Your Retirement Dollar

Your Retirement Dollar is a series of thirteen half-hour programs based on the syndicated newspaper column, "Your Retirement Dollar," by Peter Weaver, who serves as the program's host and chief expert.

The series covers a variety of topics of interest and concern to the retired, and those approaching retirement, in the areas of family finances, buying habits, good values, safe types of investments, personal pension plans, money management on a reduced income, retirement job opportunities, and so on. Two guest experts assist Mr. Weaver in further exploring subject areas. The series contains on-location footage demonstrating many of the traps facing the elderly in areas of nutrition, personal health, nursing homes, personal loans, and legal aid.

(continued)

Figure 7.1 Continued

Hollywood Television Theatre: Habit

Habit is an originally created serial for Hollywood Television Theatre. Based upon a true story, Habit explores the human condition from a uniquely sensitive view.

It is based on a profound social change that occurred as a result of the precedent-breaking Vatican Council II, and the individual struggle and change that came to a number of extraordinary women who suggested they should receive a small stipend for teaching; choose their own form of government; shift their energies from church-related to social, economic, intellectual, and spiritual needs of the family of man; and chose their own clothing to wear when working outside the College of the Immaculate Heart.

An uproar ensued, and this series tells the story in personal terms by focusing on the emotional struggles of four or five individual nuns.

The Fertile Crescent

A thirteen-part series on the history and culture of the Near East, demonstrating the accuracy of the term "Cradle of Civilization" by examining the enormous cultural contributions of the Near East to Western civilization.

It was filmed at important archaeological sites and monuments, reconstructions, and sites of custom and ritual. Also shown is the art, much of it now preserved in national museums in the Near East, in the British Museum and the Louvre, and in such fabled cities of the Near East as Baghdad, Damascus, Cairo, Jerusalem. It is safe to say that viewers will see moving sights of glory they have never beheld before.

What in the World

What in the World is a panel game show that entertains and educates by using the artifacts and treasures of the Smithsonian Institution (Museum) as its subject. The format is deceptively simple: engage two panels of three people each in a friendly, witty debate over the identity of an object from the Smithsonian.

The basic object of the game is for one panel to stump the other by weaving curious but true descriptions and stories about the object—two not fitting and one fitting the object per round—so there is a pitting of wills, wit, and intellect in the process.

Woman's Place

What does being a woman mean today? What are the special problems that our age of change thrusts into women's lives? Considering the shifts in the patterns of women's lives attending their new awakening (with responses ranging from radical feminism to Total Womanhood), what is Woman's Place—as it was, as it is, as it may become?

This series of thirteen hour-long TV programs attempts to answer these and many other questions as it presents a picture of the multiple activities and the flexible strength to be found among women who come from many backgrounds and diverse views.

Table 7.2 Evaluations of Eight New Program Concepts for Public Television by Interest Segment[a]

Program	Entire Population	Arts and Cultural Activities (AF)[b]	Cosmopolitan Self-Enrichment (M)	News and Information (M)	Highly Diversified (M)	Family-Integrated Activities (AF)	Athletic and Social Activities (Y)	Home- and Community-Centered (AF)	Mechanics and Outdoor Life (AM)	Indoor Games and Social Activities (Y)	Competitive Sports and Science/Engineering (Y)	Detached (M)	Money and Nature's Products (AM)	Elderly Concerns (AF)	Family- and Community-Centered (AM)
Just Plain Country	2.3 / 40	2.2 / 32	1.9 / 26	2.5 / 53	2.4 / 43	2.3 / 38	1.9 / 20	2.4 / 46	2.3 / 42	2.0 / 29	2.0 / 32	2.2 / 36	2.5 / 58	2.5 / 51	2.4 / 50
Mother's Little Network	2.2 / 36	2.1 / 33	2.2 / 36	2.3 / 44	2.4 / 43	2.3 / 43	2.2 / 35	2.4 / 44	2.2 / 35	2.1 / 34	2.2 / 36	1.8 / 20	1.9 / 21	2.1 / 40	2.2 / 36
Sportlight	2.3 / 44	2.4 / 47	2.3 / 46	2.8 / 61	2.7 / 61	2.0 / 32	2.4 / 41	2.1 / 40	2.2 / 37	2.1 / 30	3.1 / 78	2.0 / 28	2.4 / 51	2.0 / 29	2.6 / 52
Your Retirement Dollar	2.5 / 49	2.8 / 62	2.3 / 43	3.1 / 75	2.6 / 57	2.6 / 55	1.7 / 11	2.6 / 55	2.2 / 33	2.0 / 32	1.9 / 20	2.0 / 34	2.9 / 72	2.8 / 66	2.7 / 57
Hollywood Television Theatre	2.1 / 33	2.5 / 51	2.3 / 36	2.2 / 37	2.5 / 48	2.2 / 38	2.3 / 41	2.0 / 27	1.9 / 17	2.2 / 34	1.9 / 23	1.9 / 25	1.9 / 28	1.9 / 28	1.8 / 16
The Fertile Crescent	2.3 / 45	3.1 / 80	3.0 / 77	2.5 / 59	2.5 / 56	2.2 / 38	2.0 / 27	2.0 / 28	2.1 / 36	1.8 / 25	2.0 / 27	1.8 / 24	2.3 / 40	2.2 / 47	2.4 / 47
What in the World	2.2 / 41	2.6 / 56	2.4 / 46	2.2 / 41	2.6 / 58	2.4 / 46	2.2 / 37	2.1 / 29	2.1 / 36	2.2 / 35	2.3 / 42	1.9 / 28	2.1 / 34	2.1 / 40	2.2 / 36
Woman's Place	2.2 / 37	2.4 / 50	2.5 / 57	2.3 / 42	2.7 / 56	2.6 / 55	2.5 / 51	2.2 / 38	1.6 / 11	2.2 / 43	1.6 / 9	1.7 / 17	1.8 / 22	2.2 / 40	2.0 / 24

a. Top number in cell is mean interest rating: 4 = extremely interested, 3 = quite interested, 2 = not very interested, 1 = not at all interested. Bottom number is percentage rating 4 or 3.
b. Letters associated with each segment indicate which concentration it is associated with, namely: AF = Adult Female, AM = Adult Male, Y = Youth, and M = Mixed.

concepts for each of the five interest segments who are above average in their viewing of PTV.

Arts and Cultural Activities

Highly educated, adult women in households with manager or professional as head. Broad range of intellectual and cultural interests—especially classical arts. Low interest in household activities and management. High needs for intellectual stimulation and growth and for understanding others, with low needs for status enhancement and escape.

In general, the people in this segment are relatively interested in seeing more programming of the kinds evaluated here available on PTV. They express average or above-average interest in all program types, except for programs similar to those on commercial TV, while they are only slightly above average in their interest in "programs appealing to certain kinds of people." Consistent with their interests, they rate cultural programs and educational programs especially high, along with news and programs about local events and issues.

The Arts and Cultural Activities segment expresses above-average interest in most of the new program concepts, with members especially high in their ratings of *The Fertile Crescent, Your Retirement Dollar, What in the World,* and *Hollywood Television Theatre.* These programs, with the exception of *Your Retirement Dollar,* are laden with historical content. All appear to be intellectually enriching and serious in purpose and presentation. Concepts embodying lighter forms of entertainment attracted less interest from this segment.

Cosmopolitan Self-Enrichment

Extremely high socioeconomic profile. Diverse pattern of intellectual and cultural interests. Physically active. High needs for intellectual stimulation, unique/creative accomplishment, and understanding others. Low needs for status enhancement and for escape from boredom.

This segment's interests in types of programming for PTV mirrors that of the Arts and Cultural Activities segment, except that the latter group is more interested in news programs. Otherwise their greatest interests are in cultural programs and educational programs.

The reactions of the Cosmopolitan Self-Enrichment segment to the new program concepts, with the notable exception of *Just Plain Country*, are generally more favorable than those of the population as a whole. They, along with the Arts and Cultural Activities segment, are the only ones who rated *The Fertile Crescent* relatively high, and they showed well above-average interest in *Woman's Place*, expressing as least as much interest as the four segments with adult female concentrations. The people in this segment exhibit their lowest levels of interest in the concepts that embody light entertainment.

News and Information

Passive interests related to keeping informed on a broad range of subjects and activities. Needs are focused on being socially stimulating and maintaining family ties.

The people in the News and Information segment show their greatest interest in having more educational programs, news programs, and programs about local events and issues on PTV. These preferences are, of course, consistent not only with their interest patterns, but with their usage of commercial television and other media as well. In evaluating the program concepts, they express a higher level of interest in *Your Retirement Dollar* than does any other segment and also rate *Sportlight* relatively high. The other concepts elicited considerably less interest from them and resulted in ratings close to the population averages.

Highly Diversified

Southern, Black, adults with children. Broad range of interests, especially those permitting personal participation with family and/or other informal small group settings. High need for intellectual stimulation and growth.

The Highly Diversified segment exhibits its greatest interest in children's programs and in those program types that are heavy in news and informational content, including programs "appealing to certain kinds of people." This latter interest undoubtedly reflects the heavy concentration of Blacks in this segment. While these people are slightly above average in their ratings of all program types for PTV, they appear less interested in cultural and artistic programming (e.g., music) than in other types.

The people in this segment offered somewhat above-average ratings for every one of the eight new program concepts, with their lowest levels of interest in *Just Plain Country* and *Mother's Little Network*, the two concepts that offer the lightest entertainment.

Family-Integrated Activities

High percentage of adult women with young children. Strong interest in home and in family interactive activities—household activities and management and indoor games. High need for family ties. Child presence influences adult interest patterns.

The people in this segment are most interested in educational and children's programs and programs about special interests. For all other types of programming, they fall remarkably close to the population average in their interest in having more available on PTV.

Among the new program concepts, the highest interest for the Family-Integrated Activities segment was elicited by *Your Retirement Dollar* and by *Woman's Place*. Other ratings were close to the population average.

Summary

All of the five above-average PTV user segments discussed above produced ratings of program types for PTV and reactions to new program concepts that are consistent with their interest patterns and with their media behavior. If PTV continues to offer a mixture of cultural, informational, educational, and children's programming similar to that in existence at the time this study was conducted, these five segments are likely to remain its primary viewers. To the extent that PTV is interested in broadening its audience base and increasing PTV viewing among those people who are relatively infrequent viewers, it would appear useful to examine the reactions of the remaining nine segments to program types for PTV and to the new program concepts.

Reactions of Below-Average PTV User Segment

This section discusses the data contained in Tables 7.1 and 7.2 for each of the below-average PTV user segments in order, beginning with the Family- and Community-Centered segment, whose members are the least frequent viewers of PTV on average.

Family- and Community-Centered

Employed, blue-collar/white-collar adult males. Married, living in nonmetropolitan areas. Broad interests, including outdoor activities, investments, and home- and community-centered activities as well as religion. Very strong need for family ties.

For members of this segment, their expressed interest in having more of each of the various program types broadcast on PTV is quite similar to the population as a whole. They assign their highest ratings to educational programs, news programs, and programs about local events and issues, a pattern also observed for the News and Information segment.

The reactions to the new program concepts among members of the Family- and Community-Centered segment are neither exceptional in their range nor do they depart very much from the population averages. Their lowest ratings were for *Hollywood Television Theatre* and for *Woman's Place* (this segment is 83% male). Their highest ratings were for *Your Retirement Dollar* and *Sportlight*.

As indicated earlier, the potential for increased PTV viewing among members of this segment appears to be quite limited given their low usage of all media.

Elderly Concerns

Oldest segment, high percentage of retirees, widowed, few children. Very few interests include religion and news and information. Focus on maintaining sense of social integration and belonging in absence of direct interpersonal contact. Needs to overcome loneliness and lift spirits. Low need for intellectual stimulation.

People in the Elderly Concerns segment report an above-average interest in having more news programs and programs about local events and issues available on PTV, and below-average interest in children's programs and educational programs.

Their ratings of the new program concepts indicate that they are most interested in *Your Retirement Dollar* (with a mean of 2.8 and 66% expressing positive interest) and in *Just Plain Country*, with all other concepts rated at or below average.

As we have indicated before in discussing the members of this segment, the recurring theme that underlies the programming that appeals to them is a format that helps them to cope with loneliness and their needs for greater self-acceptance. Program content, within certain obvious limits (e.g., little interest in children's programming), is less important than a program format that tends to foster a sense of social integration by providing continuity or a mature and accepting host or leading character.

Money and Nature's Products

Older males with a higher proportion being rural and retired. Interests in passive activities that obtain some form of tangible return or product—fishing, hunting, investments. Low interest in active physical activities—camping out, participant sports—as well as culturally upscale or abstract—classical arts, international affairs. Somewhat complacent, but need interpersonal contact and support, especially from their families.

This segment was relatively discriminating in their interest ratings for program types on PTV. Their highest ratings were for news programs (2.8), where they were above the average of 2.5, programs about local events and issues, and programs about special interests. Their lowest ratings were for cultural programs,

"programs appealing to certain kinds of people," music-only programs, and children's programs.

In response to the new program concepts, members of the Money and Nature's Products segment expressed their greatest interest in *Your Retirement Dollar*, with 72% indicating a positive interest and a resulting mean rating of 2.9. Above-average, but lower, ratings were assigned to *Just Plain Country* and *Sportlight*.

The findings reinforce those reported in previous chapters and are not inconsistent with the PTV commentary in Chapter 6. In particular, these data suggest that programs dealing with practical issues, especially financial matters, or with competition could attract members of this segment into the PTV viewing audience.

Detached

> Low socioeconomic profile. Extremely few interests and activities and few psychological needs satisfied by them. Low scores on needs related to both intellectual stimulation and interpersonal contact and support.

Members of the Detached segment report levels of interest toward increased PTV offerings that are well below average for each of the nine program types rated. Their mean ratings ranged from a low of 1.6 to a high of only 2.0. Every other segment had at least one rating at or above 2.3 and most had several at higher levels.

Their mean ratings of interest in the new program concepts also fell below the population averages for every one of the concepts, ranging from a low of 1.7 for *Woman's Place* to a high of only 2.2 for *Just Plain Country*. Only the Indoor Games and Social Activities segment, a youth concentration group, approaches the Detached segment in their overall lack of enthusiasm for the new concepts. Even they, however, are somewhat more interested in most concepts than are the Detached.

The previous findings regarding these people are confirmed by these results and we continue to see no way for PTV to increase their viewing.

Competitive Sports and Science/Engineering

> Teenage male students with interests in male-associated mechanical activities and competitive athletics. Avoidance of female-oriented subjects and interests. High on needs for unique/creative accomplishment, intellectual stimulation and growth, status enhancement, and escape from boredom. Low needs for understanding others and for greater self-acceptance.

The members of this mostly male youth segment express below-average interest in all of the program types except for programs similar to those on commercial television. Their ratings are especially low for program types that are informational, educational, or culturally oriented, as opposed to those that are entertainment oriented. This result is, of course, consistent with the data previously reported on this segment's behavior with respect to television and other media as well.

In their concept ratings they report average or below-average ratings of interest in all but one concept. For that one exception, *Sportlight*, their mean interest rating of 3.1, with 78% expressing a positive interest, is far and away the most favorable response among all fourteen segments.

These results confirm the relatively narrow interest pattern of this segment and the difficulty of attracting them into the viewing audience (on a purely voluntary basis) with programming other than sports or pure entertainment.

Indoor Games and Social Activities

> Young, low-income females. Interests in activities, especially indoor games. Low interest in most subject matter areas. Nonintellectual. High needs for status enhancement and the need to be socially stimulating.

In their ratings of program types, the members of the Indoor Games and Social Activities segment express average or below-average interest in having more of almost every type available on PTV. Their interest rating for children's programs is slightly above average, reflecting the relatively large percentage of young mothers in this segment. They are especially low in their interests for news programs and programs about local events and issues.

This segment's members do not indicate notably above-average interest in any of the eight new program concepts. Their interest is understandably well below average in *Your Retirement Dollar*, and their low interest in *The Fertile Crescent* is matched only by that of the Detached segment.

Unfortunately, none of the program concepts evaluated correspond to the programming suggestions made in the PTV commentary for this segment in Chapter 6 so we do not have the opportunity to assess their reactions to these types of programs. However, there is no evidence in the data in Tables 7.1 or 7.2 to indicate that those suggestions would not be reasonably well received.

Mechanics and Outdoor Life

> Young adult, blue-collar males whose interests focus on noncompetitive activities emphasizing personal physical accomplishment—auto repair, fishing, camping—interests that, by their very nature, do not require emphasis on interpersonal cooperation or support. High on needs for escape and unique/creative accomplishment.

The people in this segment show relatively little interest in having PTV offer more of any of the program types presented to them. Their mean interest ratings ranged from a low of 1.5 for cultural programs to a high of 2.3 for programs similar to those on commercial TV and educational programs. Their interest ratings for each of the eight new program concepts are at or below average. They are especially low for *Woman's Place* (1.6 versus 2.2 average).

We suspect that this segment's interest ratings for both program types and specific concepts may be depressed by two factors. First, neither category described the types of programming that this segment is attracted to in a manner that would appeal to them. They watch escapist, action-oriented shows that are fast-moving with strong visual appeal and none of the options offered embodied these elements. Second, these people are not print oriented. Their reactions to written statements of program types and new concepts are likely to underestimate their real reactions even to programming that would be attractive to them.

Home- and Community-Centered

Adult females with a relatively high percentage of married homemakers. Home and local community interests. High needs for family ties and understanding others. Lowest needs for intellectual stimulation and for unique/creative accomplishment.

The interest ratings of the Home- and Community-Centered segment for program types mirror those of the population at large almost without exception. They are more interested in more educational and news programs than in other types and are less interested in cultural or minority programming.

Their ratings of the new program concepts are also similar to the population averages, except for an especially low interest in *The Fertile Crescent*. Their most positive rating (2.6) was for *Your Retirement Dollar*, followed by ratings of 2.4 for *Just Plain Country* and *Mother's Little Network*.

The lack of enthusiasm for the program types and new program concepts is not surprising since few, if any, are congruent with the interest patterns of these people or with their basic needs. The only exception, perhaps, is the program type, "programs about local events and issues," which received only an average interest rating of 2.4 from this segment, perhaps because of its lack of content specificity.

Athletic and Social Activities

Teenage females from high-income families. The youngest of all the segments. Interests in active, away-from-home, face-to-face activities. High need to escape from problems and to be socially stimulating. Low need for family ties.

The members of this, the youngest segment of all, exhibit their lowest level of interest in having more cultural programs available on PTV (1.6) and their highest interest (2.4) in more programs similar to those on commercial television and programs about special interests. Their interest in more news programs is the lowest of all fourteen segments.

In response to the new program concepts, people in the Athletic and Social Activities segment are most positive and well above average in their interest in *Woman's Place*, and they also rated *Sportlight* rather high. Not surprisingly, given their average age of 19, their rating of *Your Retirement Dollar* was the lowest of all the segments, as was their rating of *Just Plain Country*.

We continue to feel that the key to attracting these people into the audience for PTV lies in the coupling of educational content with an entertainment format. Many of them are still in the process of completing their formal educations and appear as a result to be resistant to taking on the burden of additional learning, unless they see it as fun.

Conclusions

The reactions of the fourteen interest segments to the expanded broadcast of selected program types on PTV and to the specific new program concepts reinforce our earlier conclusions based on actual media behavior. In earlier chapters we demonstrated the existence of a clear and logical relationship between the patterns of interests and needs that were used to structure the audience segments and the behavior of each of these segments with respect to video, audio, and print media. The data reported in the present chapter indicate that people's expressed programming preferences, as well as their actual television viewing behavior, are related to their patterns of leisure interests and needs.

The present results suggest the possibility of using the linkage between our interest segmentation scheme and their reactions to new program concepts as an inexpensive means of implementing strategies for attracting infrequent PTV viewers into the audience for PTV. Representatives of targeted interest segments could be identified by standard survey research screening procedures and asked to respond to a series of possible new program concepts designed to appeal to them, evaluating them in much greater detail, of course, than was possible in the present study. Those concepts that show the most promise could be fine tuned to maximize their appeal to their target segments and their subsequent performance on-air monitored via audience tracking studies.

Of course, not all of the people expressing interest in a program concept would watch it if it were implemented and aired. Clearly, the execution of a concept is vital to its ability to attract and retain an audience; even the manner of concept execution may be segmented in its appeal. Furthermore, interested potential viewers must be aware that a program is being broadcast, and they must be available to watch it (although availability will become less important as the number of videotape recorders in existence grows).

Nevertheless, the differential appeal of alternative new program concepts does reflect the relative *potential* for attracting audiences from the various interest segments. A concept that elicits high levels of interest has more potential, given comparable execution values, than one that elicits lower levels of interest. Similarly, other things being equal, a segment that exhibits greater interest in viewing a program based on the concept description has a greater potential to produce an audience for the program.

We believe that the diversity of ratings obtained for each new program concept across interest segments and across concepts within segments indicates that this segmentation scheme has considerable promise as a tool for integrating the early screening of new concepts into overall programming strategies for attracting target audiences.

Note

1. The new program concepts used in the survey were selected from a larger number of concepts graciously made available to the authors by the Corporation for Public Broadcasting and the Public Broadcasting Service.

8

Public Television Funding

The financial structure of PTV is complex. Its funding emanates from a variety of sources in the public and private sectors. These include the federal, as well as state and local, governments, colleges and universities, private foundations, corporations, and individuals. This diversity helps to preserve PTV's integrity and freedom from the influences and pressures of any single source.

While it is likely that PTV will continue to receive a considerable portion of its funding from congressional appropriations, and through grants and contracts from federal agencies, there is reason to believe that the 1980s will see these sources decrease as a percentage of PTV's total budget. This trend could be a healthy one if it reduces the dependency of PTV on the ever-changing political structure in Washington. During the Nixon administration in the early 1970s, PTV was seen as an adversary by the White House, and in 1972 Nixon vetoed a two-year, $155 million appropriation for the CPB that had been passed by the Congress. This led to a major restructuring of public broadcasting.

The recent Carnegie Commission (1979) argued forcefully that public broadcasting must be insulated from federal funding, both to preserve its editorial integrity, and to avoid large fluctuations in its budget resulting from the volatile political scene in Washington. In its report, the Carnegie Commission recommended that the financial support for public broadcasting derived from individual contributions be increased from $50 million in 1977 to $205 million in 1985, with $175 million directed toward PTV. Achieving an increase in individual financial support for public broadcasting, as well as PTV, requires expanding the number of individual contributions, the average amount they contribute, or, more likely, a combination of both. The chances of achieving substantial increases in funding can be improved somewhat if the program(s) developed for this purpose are based on an understanding of who contributes and why. "Program" in this context is meant to include choice of the medium used (e.g., TV, direct mail, phone solicitation, newspapers) as well as the communication content. Previous analyses of individual contributions to PTV have primarily focused on the demographic and viewing characteristics of contributors and noncontributors. Their findings have shown that contributions emanate from the more affluent viewers of culturally upscale programs.

Though the primary objective of this study was to examine the relationships between audience interest segments and media behavior, some data regarding people's past contributions and preferences for alternative funding mechanisms for PTV were obtained. These findings are reported in this chapter in the context of the fourteen interest segments. In so doing, it is hoped that this report will make a modest contribution to an improved understanding of some of the variables that impact on PTV's ability to attract contributions from its audience.

PTV Viewing and Financial Support

Table 8.1 reports the percentage of people who responded "yes" to the question, "Have you ever supported your local Public TV station with a donation?" by the frequency of their PTV viewing. Frequent viewers are defined as those who report watching programs on PTV once a week or more. They include 26% of the population. Occasional viewers (22%) are those who report watching PTV less often and the remaining group, which constitutes more than half the population, reports either never watching (47%) or not knowing if they watch or not (5%).

These data indicate quite clearly that the base of individual financial support for PTV comes from a small portion of the

FUNDING

Table 8.1 Donations to PTV by Frequency of PTV Viewing

	Entire Population	Frequency of PTV Viewing		
		Frequent	Occasional	Undetermined or Never
Percentage ever supported local PTV station with a donation	11	27	11	2

population and, as one would expect, there is a relationship between PTV viewing frequency and the likelihood of ever contributing to a local PTV station. While it is not *necessarily* true that attracting more viewers to PTV will broaden the base of contributors or increase their contributions, it is *likely* that greater involvement with PTV on the part of an individual increases his or her willingness to financially support programming efforts. This relationship between viewing and contributing also varies substantially across interest segments as reported in Table 8.2.

The last column in Table 8.2 contains an index comparing the tendency of segment members to contribute to PTV with the extent to which they view it frequently. It was computed by dividing the second column of the table by the first and converting the result to a percentage. For example, among people in the Arts and Cultural Activities segment, 29% report having ever made a voluntary donation to PTV, while 52% are frequent viewers; hence, the ratio of support to viewing is 56% (29/52 × 100).

Segments with High Support-to-Viewing Indices

The fact that the Arts and Cultural Activities and Cosmopolitan Self-Enrichment segments are by far the most frequent PTV viewers, and also the most likely to have made a donation, is not at all surprising. It confirms the lore that these two segments, who are both well above average in income, education, and cultural "taste," have long been a primary target audience for both PTV programming and fund-raising efforts.

Table 8.2 Donations to Local Public Television Station by Interest Segment (frequency in percentages)

Segment	Frequent PTV Viewers[a]	Ever Support with Donation	Ratio of Support-to-Viewing
Arts and Cultural Activities (AF)[b]	52	29	56
Cosmopolitan Self-Enrichment (M)	50	29	58
News and Information (M)	31	4	13
Highly Diversified (M)	30	12	40
Family-Integrated Activities (AF)	27	12	44
Athletic and Social Activities (Y)	25	1	4
Home- and Community-Centered (AF)	23	10	43
Mechanics and Outdoor Life (AM)	22	4	18
Indoor Games and Social Activities (Y)	21	3	14
Competitive Sports and Science/Engineering (Y)	17	9	53
Detached (M)	16	6	38
Money and Nature's Products (AM)	15	10	67
Elderly Concerns (AF)	13	3	23
Family- and Community-Centered (AM)	12	6	50
(Entire Population)	(26)	(11)	(42)

a. Percentage watching once a week or more.
b. Letters associated with each segment indicate which concentration it is in, namely: AF = Adult Female, AM = Adult Male, Y = Youth, and M = Mixed.

Equally notable, however, is the fact that 71% of the people in these two segments, including more than 40% of the frequent PTV viewers, report that they have *never* made a donation. There would appear to be an opportunity to further develop these segments' support of PTV.

Both of these segments tend to be relatively affluent and can well afford to support the PTV programming from which they derive considerable enjoyment. The viewers in these segments who have not contributed to PTV are relatively easy to reach, not only via broadcasts of theatrical and musical performances, but through print media, which they also use heavily (but selectively). They might be induced to contribute by employing some of the same fund-raising techniques used by symphony orchestras,

FUNDING

operas, art museums, and other cultural institutions. These might include, for example:

- special benefit performances of musical or theatrical events, televised and broadcast on PTV;
- discounts for substantial contributors on merchandise, which might include phonograph records or videocassette tapes of special performances, artists' prints or sculptures, specially bound books, and so forth; and/or
- publications of critical analyses and descriptions of selected musical and theatrical performances, similar to those distributed at live events. Such publications could be mailed along with PTV program schedules to "contributing members."

Three other segments have support-to-viewing indices exceeding 50%. All of them have relatively few frequent PTV viewers. The Money and Nature's Products segment, with only 15% frequent PTV viewers, claims 10% of its members as contributors to PTV. This segment, composed primarily of older adult males, is only slightly above average in income and below average in education. It does, however, include 39% retired people, compared with only 15% in the population studied. Its members may have a special appreciation for the PTV programming they do watch, however infrequently. Given their interests and values, they may feel more of a sense of obligation to financially support that from which they derive benefits. Since such a large percentage of its viewers are contributors, it appears that the key to generating more donations from the Money and Nature's Products segment is to increase the number of its members who watch PTV. A number of programming suggestions for achieving this objective were presented in Chapter 6. They include the use of themes incorporating mature, prominent, successful individuals, outdoor-related material, and the reinforcement of traditional American values.

The Competitive Sports and Science/Engineering segment also reports a relatively high percentage donating, compared to a low percentage of frequent PTV viewers. Given this segment's average age of 22 and a household income well above average, it appears likely that many of the contributions reported by members of this segment are not their own, but their parents'. Like the previous segment, increased contributions from members of this segment are also likely to depend upon attracting more of them into the PTV viewing audience. Educational programming geared to their particular interests and needs and sports programming were mentioned in Chapter 6 as possibilities for gaining viewers from this segment. Community fund-raising events, such as jogging, long distance bike riding, or marathon competitive games, could be attractive means of gaining their participation and involvement, as well as their financial support and that of their parents.

The Family- and Community-Centered segment, while containing the smallest percentage of frequent PTV viewers, yields an index of 50%. Although these people are only average in income, there appears to be a considerable potential for greater financial support from this group if they could be attracted to PTV in greater numbers. This assumes, of course, that increasing the number of frequent viewers would yield a proportionate increase in the number of donors. It would appear that religious programs and community-oriented broadcasts would be the most effective means of attracting additional viewers from this segment, although it will not be easy. Fund-raising efforts directed at this segment should stress the wholesome quality of PTV and its suitability for family entertainment and enrichment—especially in contrast to commercial television.

Segments with Medium Support-to-Viewing Indices

Four interest segments fall into the middle range, with PTV support-to-viewing indices ranging from 38% to 44%.

The Family-Integrated Activities segment ranks fifth in its percentage of frequent PTV viewers and is tied for fourth in its percentage of contributors to PTV. This segment of relatively affluent adult women with young children are especially heavy viewers of children's programs, as well as of drama programs. It is likely that the benefits they derive from PTV exceed the frequency with which they personally view it. Many members probably have young children who view PTV without their parents being present. Given the income level of this segment's members, and their usage of children's programming, there would appear to be the potential for increasing the percentage of contributors. They also represent the largest of the fourteen segments, constituting 10% of the population. Financial appeals to members of this segment should emphasize the value of PTV to the educational and social development of their children. Advertising in selected print media, such as magazines read by mothers and newspaper sections geared to young homemakers, could effectively supplement the appeals broadcast on PTV.

The Home- and Community-Centered segment, with an index of 43%, has a slightly lower percentage of frequent PTV viewers and contributors than does the previous segment. They are somewhat older, less affluent, and less educated. Their use of PTV is also less child-oriented and more likely to involve news and information. As indicated in Chapter 6, they might be attracted to watch more PTV by offering religious programming or shows that focus on local community personalities, activities and events. Like the Family-Integrated Activities segment, they can be reached through selected print media directed at homemakers. The most effective appeals should probably focus on the value of PTV to the development of their children.

The Highly Diversified segment has the fourth highest percentage of frequent PTV viewers and is tied for third in the percentage contributing to local PTV stations. As one of the least affluent segments, this segment, perhaps more than any other, illustrates the potential for broadening the base of financial support for PTV, providing that audience interest and commitment can be generated. Because the members of this segment tend to have relatively low household incomes, they are not likely to be in a position to make substantial financial contributions to PTV. However, since they do appear to derive what they see as important benefits from PTV, they might well respond to the kind of financial appeal that solicits broad participation, no matter how small the amount. A broad range of themes should appeal to members of the Highly Diversified segment, as they use PTV for their own educational and cultural development, as well as for their children's, and see it as a means of family integration.

The fourth segment in this middle range is the *Detached* group. They are low in PTV viewing and in the percentage contributing to PTV. Given their pattern of interests and needs as discussed in Chapter 3, it is not likely that there is much potential for increased contributions from these people.

Segments with Low Support-to-Viewing Indices

The remaining five segments all contain people with relatively low ratios of financial support-to-viewing frequencies. It is noteworthy that not all of them represent infrequent PTV viewers, but that they are rather evenly distributed across the range of the fourteen segments.

The lowest index of all (4%) is associated with members of the Athletic and Social Activities segment. While 25% of them report frequent viewing of PTV, only 1% report ever contributing a donation. This result undoubtedly reflects the fact that they are primarily adolescent girls, averaging 19 years of age. The potential for increased contributions among them is probably quite low over the near term. However, they do come from relatively affluent households and may well become contributors later in life.

Next lowest is the News and Information segment, which is well above average with 31% frequently watching PTV, but only 4% reporting ever having made a donation. They rank third in usage, by this measure, but tenth in contributions. It would appear that the PTV viewing of this segment, more than any other, has been heavily subsidized by the contributions of others. They are about average in income. Yet for some reason, they have not responded well to PTV's fund-raising efforts. To the extent that the ability to generate public financial support is employed as a criterion for selecting target audiences, this segment would be attractive only if methods can be devised for reaching them with messages that would persuade them to contribute more than they have in the past. Clearly they can be reached with messages placed in print, radio, or television associated with the reporting of news.

The Indoor Games and Social Activities segment, the other group of young women (average age is 22), also exhibits an extremely low proportion of members (3%) who have ever made a contribution to PTV. Given their youth and low-income status at this stage in their lives, they show little potential for increased support of PTV at this time.

The Mechanics and Outdoor Life segment is only slightly below average in their percentage of frequent PTV viewers, but well below average in their percentage of PTV contributors. There is no PTV program to which they are attracted in large numbers, and apparently their level of commitment to PTV is minimal.

Finally, the Elderly Concerns segment has extremely few frequent viewers of PTV and few contributors. As the segment with by far the lowest income, the potential for increased financial contributions from them appears to be small.

The data reported above clearly indicate that the relationship between PTV viewing behavior and financial support of local PTV stations is not a simple one. While our data are not sufficiently detailed to permit a comprehensive economic analysis,[1] it is clear that viewers of PTV in some interest segments are more likely to contribute than viewers in other segments. Furthermore, this likelihood is not a simple function of household income, but appears to be linked to involvement with PTV and commitment to it.

Preferred Sources of Funding for PTV

A number of alternative methods for funding PTV have been considered by the Carnegie Commission, the federal government, the Corporation for Public Broadcasting, and by other interested parties. The public's preferences as to principal funding alternatives were measured by asking each person to choose which one of the following they would most like to see as the major source of funding for PTV:

(1) use of federal taxes to fund public TV
(2) state government funding of public TV
(3) local community funds for public TV

Table 8.3 Preferred Sources of Funding by Frequency of PTV Viewing (in percentages)

Funding Source	Entire Population	Frequency of PTV Viewing		
		Frequent	Occasional	Undetermined or Never
Commericals	30	17	31	36
Income tax checkoff	17	24	14	14
Federal taxes	13	14	15	12
Networks taxed	12	13	11	12
Local community funds	11	15	10	10
State government	10	12	13	8
TV set tax	6	8	6	5

(4) television set tax—a tax placed on each new TV set purchased—revenue from this tax would be used to support public TV
(5) national networks taxed—commercial television networks would be taxed and these taxes would support public TV
(6) commercial time sold *between* public TV programs to support public TV
(7) income tax checkoff—every individual would be given an option to designate a small portion of their income tax payment to be used for the support of public TV

Table 8.3 presents the responses to this question as a function of frequency of PTV viewing.

Overall, commercial time sold between programs was by far the most popular choice. It is likely that most people, while often irritated by commercials, do not perceive them as involving any cost to themselves, the viewers. This is also the most familiar form of subsidizing television broadcasts to the majority of the population, who still do not pay directly for the television brought into their homes. The second most preferred was an income tax checkoff, with a tax on TV sets being the least preferred.

Among frequent PTV viewers there was a clear reversal in the first two preferences, with an income tax checkoff being most preferred, and the showing of commercials between programs ranking second. Since the frequent PTV viewer is the one most impacted by the showing of commercials, this reversal is not at all surprising. With this single exception, there appears to be little, if any, relationship between frequency of PTV viewing and preferred sources of funding.

Table 8.4 reports the same data for each of the fourteen interest segments.

Above-Average PTV User Segments

People in the Cosmopolitan Self-Enrichment segment are least favorably inclined toward the selling of commercial time on PTV, strongly preferring the implementation of an income tax check-off, or even local community funding. The Arts and Cultural Activities segment is split between commercials and an income tax checkoff, with other preferences spread evenly over the remaining choices. The News and Information segment is notable as having the highest percentages among all fourteen segments supporting state government funding and for a tax on the networks. The Highly Diversified segment offers the strongest support for the use of federal taxes, but still prefers commercials. Finally, the Family-Integrated Activities segment, among the above-average PTV viewers, exhibits the largest percentage favoring the use of commercials.

Below-Average PTV User Segments

Among these nine segments the preference for commercials as the major funding source is consistently strong with some variations in their rankings of the remaining six alternatives. The income tax checkoff remains the second most frequently preferred for four of these segments, but is not at all popular among the Athletic and Social Activities and Family and Community Centered segments.

Conclusions

As indicated in the introduction to this chapter, the financial structure of public broadcasting and PTV has been both complex and subject to dramatic, and often adverse, changes in funding,

Table 8.4 Preferred Sources of Funding by Interest Segment[a] (in percentages)

Funding Source	Entire Population (average)	Arts and Cultural Activities (AF)[b]	Cosmopolitan Self-Enrichment (M)	News and Information (M)	Highly Diversified (M)	Family-Integrated Activities (AF)	Athletic and Social Activities (Y)	Home- and Community-Centered (AF)	Mechanics and Outdoor Life (AM)	Indoor Games and Social Activities (Y)	Competitive Sports and Science/Engineering (Y)	Detached (M)	Money and Nature's Products (AM)	Elderly Concerns (AF)	Family- and Community-Centered (AM)
Commercials	30	26	16	26	30	35	28	37	32	37	34	30	27	31	33
Income tax checkoff	17	26	32	16	9	15	9	12	24	19	16	10	19	14	9
Federal taxes	13	10	10	14	21	14	14	13	10	7	12	14	17	14	14
Networks taxed	12	12	10	18	9	12	14	9	10	10	14	10	14	17	15
Local community funds	11	16	19	8	13	6	14	9	12	10	11	17	9	6	7
State government funding	10	11	7	18	11	13	12	12	7	8	11	7	7	5	11
TV set tax	6	5	10	2	9	4	7	7	5	6	4	7	7	6	8

a. Some column percentages add to less than 100% due to "no answers" and "don't knows," while others add to greater than 100% due to multiple responses.
b. Letters associated with each segment indicate which concentration it is in, namely: AF = Adult Female, AM = Adult Male, Y = Youth, and M = Mixed.

depending on the administration in Washington. This situation is likely to continue as long as the federal government remains the primary source of revenue.

To further complicate matters, the environment in which PTV competes for viewers is changing as well. People who used to view television broadcasts as a "free commodity," paid for by the networks and by commercials, are now spending increasing amounts of money for programming brought into their homes via cable, satellite, VCR, and videodiscs. Already there are indications that the commercial networks and cable programmers are exploring ways to target arts and cultural programming against the influential, acquisitive, and affluent people who are known to be attracted to such broadcasts. Many of these fall into the Cosmopolitan Self-Enrichment and Arts and Cultural Activities segments that currently constitute a significant portion of the audience for PTV. As a result, PTV may face serious competition for the first time in a domain where it has held an exclusive franchise for most of its history.

Given this environment, it is quite likely that PTV will have to seek greater financial support directly from its viewers. Such support could come from voluntary contributions and auctions as it does now, or it could be derived from a pay channel. In either of these cases it would become increasingly necessary for PTV to develop strategies for program development and broadcast that will attract audiences willing to financially support such programming.

This chapter has attempted to present a conceptual framework for examining strategies to be explored by PTV in such an environment. It appears that some segments that have been among the above-average viewers of PTV have been unwilling to support it financially (e.g., the News and Information segment). In the environment of the 1980s can PTV afford to continue its appeal to this audience? Should it? Who will pay the bill? Similar questions will arise regarding other audience segments. As PTV changes, its appeal to the various interest segments will change. Some who are heavy viewers today will view less and vice versa. What will not change is the increasing competition that PTV will face to attract viewers in every segment. These issues, while couched in economic terms, are much more far reaching, of course. They raise challenges to the fundamental political and social philosophy that led to the formation of public broadcasting and will have to be addressed within that context.

Note

1. We do not, for example, know either the frequency with which contributions have been made, nor the size of the contributions.

9

Minority Audiences

The attraction of minority audiences has long been one of the expressed goals of public broadcasting in general, and public television in particular. This goal has developed partly in response to PTV's basic mission of offering a diversity of high quality programming of a type not offered by commercial television. Additionally, its heavy dependence upon federal funding has subjected PTV to increasing pressures from minority groups to serve their memberships more directly.

The 1960s and 1970s saw the emergence of Blacks as a vocal and powerful political group within the United States. Late in the 1970s Hispanic Americans, particularly those concentrated in Florida, the Southwest, Southern California and New York, began to attract attention. Finally, the gradual increase in the median age in this country leads to the inevitable conclusion that the next two decades will see a dramatic increase in the commercial influence of the elderly population on the mass media.

Each of these minority groups, Blacks, Hispanics, and the elderly, has in recent years become a target for commercial, as well as political, interests. Manufacturers and advertisers of consumer products and services have recognized their current and potential buying power. In response, they are seeking means of developing and marketing new offerings designed to meet their needs.

As a consequence, during the 1980s and beyond, PTV is likely to encounter increasing competition from commercial programming and broadcasting interests for the viewing time of these and other minority audiences. If PTV is to attract larger numbers of minority viewers, it is likely that new strategies for doing so will need to be explored as a result of the changing competitive environment.

Our interest segmentation scheme provides one conceptual framework for developing such strategies. If one's sole purpose were to attract minority audiences, one would probably study the particular minority groups of interest, examine their behavior and needs, and develop programming similar to what they currently view, along with new shows designed to meet their unfulfilled needs. Given the complexity of PTV's overall objectives, particularly the desire to attract nonminority as well as minority audiences, and to offer a quality and type of programming not otherwise available, we believe that a more balanced approach can be taken. Minority audiences are distributed across all of our fourteen interest segments. By targeting programming against interest segments that offer high potential for minority viewing, PTV should be able to increase the size of its minority audiences. Blacks can be attracted not only by offering "Black programming" in the narrow sense, but also programming that appeals to the general interests and needs of the segments that contain a relatively large proportion of Blacks. Other minorities can be attracted in a similar manner.

This chapter focuses first on the distribution of three key minority groups—Blacks, Hispanics, and the elderly—across the fourteen interest segments, identifying segments that could serve as strategic targets for increased minority viewing. Second, we explore some of the general characteristics that distinguish minority viewers from the general population as a means of providing a sharper focus on how to attract the minority viewers within these segments.[1]

Minority Composition of Interest Segments

Blacks were self-identified by their responses, to a question on race and, after appropriate weighting, were estimated to represent 11% of our entire population.[2] Those classified as Hispanics

Table 9.1 Minority Composition of Interest Segments (in percentages)

Interest Segment	Entire Population	Black	Hispanic	Elderly
Adult Male Concentration				
Mechanics and Outdoor Life	8	2	5	2
Money and Nature's Products	6	4	4	13
Family- and Community-Centered	6	4	3	9
Adult Female Concentration				
Elderly Concerns	8	9	5	26
Arts and Cultural Activities	9	8	8	10
Home- and Community-Centered	8	8	9	9
Family-Integrated Activities	10	4	8	4
Youth Concentration				
Competitive Sports and Science/Engineering	6	5	11	*
Athletic and Social Activities	4	2	8	—
Indoor Games and Social Activities	4	8	5	—
Mixed				
News and Information	5	6	3	7
Detached	9	16	13	14
Cosmopolitan Self-Enrichment	8	1	6	2
Highly Diversified	8	24	13	4
Approximate Sample Size	(2476)	(529)	(345)	(405)

*Less than 0.5%.

responded to the question on race by identifying themselves as "Spanish (such as: Central or South American, Cuban, Chicano, Mexican, Mexican-American, Puerto Rican, other Spanish)." As reported in Chapter 2 the large majority of Hispanics were not a part of our national probability sample, but were drawn from fifteen areas known to have high concentrations of Hispanics. To have attempted to locate a sufficiently large sample of Hispanics within our basic national probability sample frame would have been inordinately costly given the low incidence of Hispanics in most areas of the United States. The 405 people classified as elderly are those individuals drawn from our national sample who reported themselves to be age 50 or older. Table 9.1 reports the distributions of our three minority populations across the fourteen interest segments.

Blacks

The Black population ranges from a low of 1% in the Cosmopolitan Self-Enrichment segment to a high of 24% in the Highly Diversified segment, while the population as a whole ranges from a low of 4% in two of the Youth Concentration segments (both primarily female) to a high of 10% in the Family-Integrated Activities segment.

Besides the Highly Diversified segment, Blacks exist in substantial numbers in the Detached (16%), Elderly Concerns (9%), Arts and Cultural Activities (8%), Home- and Community-Centered (8%), and Indoor Games and Social Activities (8%) segments. Only two of these, the Highly Diversified and the Arts and Cultural Activities segments, are among the five that are currently above average in their viewing of PTV as reported in Chapter 5.

Hispanics

Hispanics are dispersed much more evenly across the interest segments than are the Blacks, although there is still considerably more variation than in the population as a whole. At the low end is the Hispanic membership in the Family- and Community-Centered (3%) and News and Information (3%) segments. The segments containing the largest proportions of Hispanics are the Highly Diversified (13%), Detached (13%), Competitive Sports and Science/Engineering (11%), Home- and Community-Centered (9%), Arts and Cultural Activities (8%), Family-Integrated Activities (8%), and Athletic and Social Activities (8%) groups.

Three of these segments, with relatively large numbers of Hispanics, namely, the Highly Diversified, Home- and Community-Centered, and Arts and Cultural Activities segments, are among the five that are above average in their PTV viewing.

Table 9.2 Minorities in Segments with Above-Average PTV Viewing (in percentages)

Segment	Entire Population	Black	Hispanic	Elderly
Cosmopolitan Self-Enrichment (M)	8	1	6	2
Arts and Cultural Activities (AF)	9	8	8	10
News and Information (Y)	5	6	3	7
Highly Diversified (M)	8	24	13	4
Family Integrated Activities (AF)	10	4	8	4
Total	40	43	38	27

Elderly

The distribution of the elderly among our fourteen interest segments is clearly the most variable of the minority groups, ranging from zero (two of the Youth Concentration segments failed to include a single elderly person) to 26% in the Elderly Concerns segment. In addition to this segment, relatively large numbers are found in the Detached (14%), Money and Nature's Products (13%), Arts and Cultural Activities (10%), Family- and Community-Centered (9%), and Home- and Community-Centered (9%) segments. Of these six segments that comprise 81% of the elderly population, only one, the Arts and Cultural Activities segment, is among those that are currently above average in their viewing of PTV.

Table 9.2 compares the minority composition of the five segments with above-average PTV viewing with that of the general population. It is interesting to note that, while Blacks and Hispanics are underrepresented in the Cosmopolitan Self-Enrichment and Family-Integrated Activities segments, their overrepresentation in the Highly Diversified segment causes their total membership across all five segments to be close to that of the general population. The elderly, on the other hand, are clearly underrepresented among the relatively heavy PTV viewing segments.

Target Segments for
Increased Minority Viewing of PTV

We have selected five of the fourteen interest segments that appear to offer the greatest potential for attracting larger numbers of minority viewers into the audience for PTV. They are:

- Highly Diversified
- Family-Integrated Activities
- Home- and Community-Centered
- Elderly Concerns
- Money and Nature's Products

The first two of these target segments were drawn from segments that are currently above average in their viewing of PTV, while the latter three are below average.

Above-Average PTV Viewing Segments

The segments to be targeted were chosen with several criteria in mind. First, each of the selected segments contains an appreciable number of one or more of the three minority groups. Second, each indicates through their current patterns of interests and needs, and their television viewing behavior, the potential for viewing more PTV than they are currently watching. A third criterion draws on the data presented in Chapter 7, reporting their expressed interest in various types of programs for PTV and in a sampling of new program concepts for PTV.

Among the five above-average PTV viewing segments, the Highly Diversified and the Family-Integrated Activities segments appear to offer the greatest potential for increased PTV viewing by minorities. The Highly Diversified segment comprises extremely large numbers of Blacks and Hispanics and, while they are currently relatively heavy viewers of PTV, their appetites do not appear to be satiated. The Family-Integrated Activities segment is the largest of all, with 10% of the total population, and contains 8% of the Hispanics surveyed.

While the Arts and Cultural Activities segment includes substantial numbers of all three minorities, the members of this segment are already watching a lot of PTV, and the potential for increased viewing may be quite limited. The Cosmopolitan Self-Enrichment segment contains very few minority members, as does the News and Information segment, which only comprises 5% of the total population to begin with.

Below-Average PTV Viewing Segments

All three minorities are well represented in the Home- and Community-Centered segment. Furthermore, this segment contains an above-average number of daily PTV viewers, despite the fact that 49% of its members report never watching it. The Money and Nature's Products segment contains 13% of the elderly population, primarily men, and the Elderly Concerns segment contains 26% of the elderly population, primarily women, and 9% of the Black population. These two segments provide the primary opportunities for attracting the elderly into the PTV viewing audience. While the Detached segment includes above-average numbers of all three minorities, their patterns of interests and needs make it difficult to develop strategies for involving them, and so they are not included as a target segment.

Other segments, such as those with large concentrations of youths, were not targeted either because they compromise relatively small numbers of minority members or because there appear to be few opportunities to attract viewers in these segments into the PTV audience.

Comparisons of Minorities with the General Population

Having identified five interest segments as targets for increased minority viewing of PTV, the development of strategies for attracting minorities involves two components. The first is directed at attracting each of the target interest segments as a whole. Suggestions for achieving this end were presented in Chapters 5 and 6. The second is directed at the minority groups *within* each of these target segments to ensure that they are attracted at least in proportion to their membership in the segment, and perhaps in even greater proportion. Toward this end, this section compares the demographic characteristics, the current PTV viewing behavior, and interest in PTV program types and new program concepts of minorities with the general population.[3] The concluding section of this chapter incorporates these comparisons in a discussion of strategies for attracting minorities in each of the five targeted interest segments into the audience for PTV.

Demographics

Table 9.3 compares the demographic characteristics of the Black, Hispanic, and elderly populations with those of the entire population. In examining these data, it is important to remember that the Blacks and elderly were sampled on a strict area probability basis and that their data are projectable to the population as a whole. The Hispanics were drawn from selected areas on a quota basis and do not necessarily represent the population of Hispanics in the United States.

Among Blacks, the distribution of males and females conforms pretty well to that of the population as a whole. Blacks are somewhat younger and less likely to be married. They are much more likely to reside in central-city areas and are heavily overrepresented in the South. Their mean income is well below the population average, primarily, it appears, due to their underpresentation in the managerial/professional occupations and their lower level of education.

The Hispanics also mirror the population's ratio of males to females; they tend to be much younger on average than the population in general and to have more children under age 13. The Hispanic sample was drawn primarily from central cities in the West and Northeast and has a much lower income than the population average. The members of this sample are also less well educated and are underrepresented in the managerial/professional category.

The higher percentage of women among the elderly population undoubtedly reflects their greater life expectancy. Almost all are

Table 9.3 Demographic Comparison of Total and Minority Samples (in percentages)

Characteristic	Entire Population	Blacks	Hispanics	Elderly
Sex				
Male	48	46	47	43
Female	52	54	53	57
Age				
13 to 17 years	14	19	23	—
18 to 34 years	33	36	38	—
35 to 49 years	21	20	23	—
50 years and over	32	26	16	100
Mean (years)	40	36	32	69
Marital Status				
Single	22	33	35	2
Married	64	48	54	63
Divorced, separated	6	10	9	3
Widowed	8	9	2	32
Adults with Children				
(under 13 years)	46	46	57	10
Average age of children (years)	11	11	11	14
Urban/Nonurban				
Central city	24	48	76	27
Suburb	41	24	15	32
Nonmetro	34	28	9	41
Region				
Northeast	24	11	28	24
Central	24	20	8	25
South	35	63	13	34
West	17	6	50	17
Mean Annual Income				
(in thousands of dollars)	14.2	9.0	10.5	8.4
Employment Status				
Employed full time	40	38	34	11
Employed part time	7	6	7	6
Temporarily unemployed	5	7	6	1
Retired	15	11	5	62
Occupation[a]				
Blue collar	25	25	25	31
White collar	21	25	18	24
Managerial/professional	20	12	9	23
Homemaker	19	15	23	22
Student	15	24	25	—
Education				
Grammar school only	12	18	33	28
High school only	55	52	49	45
College	33	26	18	27
Approximate n =	(2476)	(529)	(345)	(405)

a. Includes previous occupations for those retired or temporarily unemployed.

Table 9.4 Total and Minority Viewing of PTV (in percentages)

Frequency	Entire Population	Blacks	Hispanics	Elderly
Every day	7	13	6	5
One to six times a week	19	18	20	18
Once or twice a month	10	5	7	7
Once to "a few" times a year	12	12	16	5
Never	47	46	45	58
Don't know	5	8	6	6
Approximate sample size =	(2476)	(529)	(345)	(405)

either married or widowed and, of course, few have young children living at home with them. The elderly population is dispersed across the United States in proportion to the rest of the people, but they are more likely to reside in nonmetropolitan or rural locations and less likely to live in the suburbs. Their average household income is even lower than that of Blacks or Hispanics. This is largely because 62% are retired compared with 15% in the population as a whole. Their level of formal education is below average with 28% not going beyond grammar school.

Frequency of PTV Viewing

Our data, as presented in Table 9.4, indicate that Blacks and Hispanics watch PTV as much as, or perhaps even more than, the population as a whole. These results, based on self-reported data, appear somewhat discrepant with data reported by Myrick and Keegan (1981) for the Corporation for Public Broadcasting. They report Nielsen data indicating that, in 1979 during a one-month period, PTV reached 58.8% of the nonwhite households in the country and 66.6% of the white households. They also report other sources of data indicating that nonwhites are slightly underrepresented among households viewing PTV. Given the substantial differences between data collected on a recall basis and diary or metered data, and between individual viewing and household viewing data, and between overall and one-month usage data, these discrepancies are understandable. For example, one would expect higher percentages for household viewing data than for individual viewing data. One would also expect that the four-week recall method used in the present study would be more likely to understate actual viewing than would either the diary or meter method. It is not at all clear how such differences might vary across minority groups, but the potential is certainly there.

Our data indicate that there are no major barriers to Black and Hispanic viewing of PTV and that, in fact, the percentage of Blacks viewing it every day (13%) is almost twice that of the population as a whole (7%). The reported pattern of viewing frequency for Hispanics mirrors that of the general population. The elderly, however, are clearly underrepresented among PTV viewers, with 64% claiming either that they never watch it or don't know if they do or not.

Types of Programs Viewed

The first column of Table 9.5 indicates the average monthly viewing frequency for each of 19 types of programs for the population as a whole. The remaining columns report comparable data in the form of a percentaged index for each of the three minority groups. Thus, the Black index of 154% for adventure programs indicates that Blacks reported viewing 54% more programs in this category during the past four weeks than did the population as a whole.

These data indicate that Blacks tend to be substantially heavier viewers of virtually all types of television shows except daily news shows and documentaries. While situation comedies and crime dramas are the program types viewed most frequently by the general population, Blacks tend to view both types at rates more than 50% above average. Especially popular among Blacks are musical performances (a category that included programs such as *Soul Train*), science fiction, other (which includes religious programs), and soap operas.

Additional data indicate that at the individual program level, Blacks are especially heavy viewers of programs in which the main characters or themes are Black. For example, their viewing

Table 9.5 Total and Minority Average Monthly Viewing Frequency of Program Types

	Total Frequency	Black Index[a]	Hispanic Index[a]	Elderly Index[a]
Adventures	4.14	154	123	104
Children's programs	1.36	162	173	24
Crime dramas	7.70	159	135	106
Documentaries	.34	106	109	132
Dramas	4.42	114	89	135
Game shows	3.56	153	97	140
Movies	5.45	136	134	76
Musical performances	1.02	274	127	92
News/commentaries	2.37	124	46	173
News shows—daily	5.19	99	81	134
Science fiction	1.25	209	155	59
Situation comedies	19.15	151	118	92
Soap operas	3.64	208	111	139
Specials	.89	131	75	88
Sports	5.79	138	103	90
Talk shows	2.90	116	81	152
Theatrical performances	.63	111	62	171
Variety shows	3.63	174	114	147
Other	.67	203	200	206
Approximate n =	(2476)	(529)	(345)	(405)

a. Ratio of minority group frequency to total frequency × 100.

indices for several such programs include *Soul Train* (630%), *Black Perspective on the News* (643%), *Sanford and Son* (486%), *The Richard Pryor Show* (464%), *Redd Foxx* (311%), *Good Times* (288%), and *What's Happening* (249%).

Hispanics appear to be lighter users of television in general than are Blacks. They are relatively frequent viewers of the "other" category (which included any Spanish programs), variety shows, children's programs, and science fiction programs, and are moderately frequent viewers of such "action-oriented" types as crime dramas, movies, and adventures, along with musical performances. Hispanics are well below average in their viewing of news/commentaries, daily news, theatrical performances, talk shows, and specials. All of these tend to rely heavily on verbal presentations. The less frequent viewing of these program types by Hispanics may be related to a lower level of facility with the English language or to their generally lower level of education.

The elderly are relatively heavy viewers of the "other" category, primarily because it includes religious programs. They are also above-average viewers of news/commentaries, daily news, theatrical performances, variety shows, talk shows, game shows, soap operas, dramas, and documentaries. They clearly watch more daytime television than average, in part because they are available at home to watch it. Their evening viewing is selective. They are relatively light viewers, for example, of movies, science fiction, specials, sports, musical performances and even situation comedies.

Interest in Program Types for PTV

Table 9.6 presents data showing how interested members of minority groups said they would be in being able to see more of certain types of programs in the future on PTV.

Blacks appear to look toward PTV as a source of information and education. The types of shows they would most welcome in greater frequency on PTV are: educational programs, programs about local events and issues, news programs, children's programs, and "programs appealing to certain kinds of people (i.e., women, Blacks, Spanish-speaking, older people, etc.)." More than half the Black respondents said they would be "somewhat" to "very" interested in seeing each of these types of shows more often on PTV. Compared to the total population, they are notably more interested in "programs appealing to certain kinds of people," children's programs, and programs about local events and issues. Blacks are notably less interested in "programs about special interests (i.e., home gardening, tennis, the stock market, etc.)."

Like other viewers, Hispanics report that they are especially interested in seeing more educational and news programs on PTV. Hispanic interest in additional news shows, despite infrequent news viewing among Hispanics currently, suggests that Hispanics have certain special needs in terms of news coverage

Table 9.6 Total and Minority Interest in Seeing More of Certain Program Types on PTV[a] (in percentages)

	Entire Population	Blacks	Hispanics	Elderly
Cultural programs	34	32	31	38
(mean rating)[b]	(2.0)	(1.9)	(2.0)	(2.1)
Programs similar to those on commercial TV	42	46	47	38
(mean rating)	(2.2)	(2.2)	(2.3)	(2.1)
Music-only programs	36	40	44	38
(mean rating)	(2.0)	(2.1)	(2.2)	(2.1)
Programs appealing to certain kinds of people	28	53	50	26
(mean rating)	(1.9)	(2.5)	(2.4)	(1.8)
Programs about local events and issues	55	60	47	68
(mean rating)	(2.4)	(2.5)	(2.3)	(2.6)
News programs	56	57	54	72
(mean rating)	(2.5)	(2.6)	(2.5)	(2.8)
Programs about special interests	52	40	38	52
(mean rating)	(2.4)	(2.2)	(2.2)	(2.4)
Educational programs	61	62	58	56
(mean rating)	(2.6)	(2.7)	(2.6)	(2.5)
Children's programs	47	56	45	28
(mean rating)	(2.3)	(2.6)	(2.4)	(1.9)
Approximate n =	(2476)	(529)	(345)	(405)

a. Percentage saying they would be "very" or "somewhat" interested in more of each type of program.
b. Based on a 4-point scale for each program category: 1 = "not at all interested" and 4 = "very interested."

that are not being met. Programs focusing on special interests elicit relatively little interest among Hispanics, but "programs appealing to certain kinds of people" are of much greater interest to them than to the population in general. This probably reflects a desire to see more programs with Hispanic characters, themes related to Hispanic life, or programs with Spanish-language subtitles. Other types generating above-average interest include

Table 9.7 Total and Minority Interest in New Program Concepts for PTV[a] (in percentages)

Program Concept	Entire Population	Blacks	Hispanics	Elderly
Just Plain Country	40	26	34	48
(mean rating)[b]	(2.2)	(1.9)	(2.2)	(2.5)
Mother's Little Network	36	39	41	36
(mean rating)	(2.2)	(2.2)	(2.2)	(2.1)
Sportlight	44	49	54	37
(mean rating)	(2.3)	(2.5)	(2.6)	(2.1)
Your Retirement Dollar	49	56	47	74
(mean rating)	(2.5)	(2.5)	(2.4)	(3.1)
Hollywood Television Theatre—Habit	33	46	48	28
(mean rating)	(2.1)	(2.4)	(2.4)	(2.0)
The Fertile Crescent	44	41	47	48
(mean rating)	(2.3)	(2.2)	(2.4)	(2.4)
What in the World	41	43	43	42
(mean rating)	(2.2)	(2.3)	(2.4)	(2.2)
Woman's Place	37	58	45	37
(mean rating)	(2.2)	(2.6)	(2.4)	(2.1)
Approximate n =	(2476)	(529)	(345)	(405)

a. Percentage saying they would be "quite" or "extremely" interested.
b. From 4-point rating scale: 1 = "not at all interested" and 4 = "extremely interested."

"programs similar to those on commercial TV" and music-only programs.

Clearly, the elderly value television—or at least PTV—as a source of information. They especially want PTV to provide more news programs and more programs about local events and issues. Outside of the elderly group, this level of interest in these two types of programs is unmatched, and inside the group it far exceeds interest in any other type of show.

Interest in New Program Concepts

Table 9.7 compares the interests expressed by the three minorities in each of the eight new program concepts for PTV as described in Chapter 7.

Again, the need for sources of information and guidance in coping with a changing social and economic environment is suggested by Blacks' interest ratings of these verbal descriptions of possible new shows for television. The highest percentages of "quite" or "extremely" interested ratings were accorded to *Woman's Place,* an exploration of the role of women in today's society, and *Your Retirement Dollar,* a practical program in financial planning. Black interest levels exceeded those of the general population in another concept, *Habit,* involving a controversial social issue (dealing with a recent conflict within the Catholic church) and on *Sportlight.* Blacks expressed considerably less interest in *Just Plain Country,* featuring country music, than did either of the other minority groups or the general population.

The Hispanic group expressed considerably higher interest than the population at large in *Sportlight* and in *Habit.* The latter program may have been especially appealing because so many of the Hispanic group are Catholics. Other concepts generating almost 50% responses of "quite" or "extremely" interested were the more traditional educational/informational prototypes—especially *Your Retirement Dollar, The Fertile Crescent* (a historical-archaeological program), and *Woman's Place.*

High interest among the elderly in *Your Retirement Dollar* might be expected, not only because of their age and retirement status for many of them, but also because of their professed preference for informational, nonfictional programming. Almost three-fourths of this group said they would be "quite" or "extremely" interested in such a program. We need not conclude, however, that any new program must be either informational or narrowly focused on elderly concerns in order to appeal to elderly viewers. Relatively high levels of interest were also expressed in *Just Plain Country* and in *The Fertile Crescent.* It is not clear why they exhibit above-average interest in *The Fertile Crescent,* given the generally low need for intellectual stimulation reflected among most of the elderly population. Perhaps the particular nature of this concept appeals to a sense of history that often characterizes the elderly.

Conclusions

Earlier in this chapter we identified five interest segments as primary targets for attracting more minority members into the viewing audience for PTV. Having compared a number of characteristics of minorities with each other, and with the population in general, we are now prepared to suggest strategies for attracting the minority members within each of those interest segments into the PTV audience. These strategies are based upon our knowledge of the interests and needs of each of these segments combined with what we know about the Black, Hispanic, and elderly populations in general.

Highly Diversified

Because of the diverse interests of this segment's members, there are a variety of ways to attract them into the viewing audience. They express above-average interest in educational, news, children's, and "special interest" programs, and in almost every other kind of program offered. Because these people are also heavy viewers of commercial television, PTV is in constant competition with commercial broadcasters for their attraction.

It would appear that increased PTV viewing among this segment could be obtained by offering programs that encourage family viewing and combine the elements of entertainment with intellectual stimulation and growth. The key elements of such programming might include a serious (but not necessarily heavy) dramatic treatment of family life in which Blacks and/or Hispanics are portrayed in strong roles. Since this segment watches such a wide variety of television, it is especially important to schedule programs targeted toward reaching them at times when they are likely to be easily diverted from commercial offerings.

Creating awareness of programs can be achieved through several media. This segment uses radio very heavily (including public radio), has moderate readership of newspapers, and exceptionally heavy magazine readership—especially Black magazines.

They also rely heavily on *TV Guide* as a source of information about television.

Family-Integrated Activities

The Family-Integrated Activities segment includes a relatively high percentage of daily PTV viewers and a large percentage of nonviewers. Consistent with their family orientation, members of this segment report especially high interest levels in seeing educational and children's programming as well as special interest programming on PTV. Members of this segment appear to use television as a means of furthering the intellectual and social development of young children, and they often view television with their children as a means of bringing the family together and strengthening family ties.

Clearly, the Hispanics in this segment would be especially responsive to Spanish-language broadcasts or to programs with Spanish-language subtitles. They could be attracted not only to programs with a clear educational mission, but also to nonviolent dramas with a family focus. Such program types have attracted them to both commercial broadcasting and PTV in the past, and appear to have substantial potential for significantly increasing their viewing of PTV.

Programming directed at this segment can be advertised and promoted via radio and print media. This segment shows exceptionally high usage of radio during weekdays (pop, rock or top hits) and relatively high newspaper readership. They are light in news readership, but heavy in the same types of features that appeal to the Home- and Community-Centered segment: gardening, cooking, social news, advertising, and personal advice. They have the highest readership of women's service magazines of any segment.

Home- and Community-Centered

The Home- and Community-Centered segment is interested in news and local topics, but also shows the potential to be attracted to educational and specialized interest programs. It appears that this segment could be attracted to PTV by programming addressed to young children or by programming that satisfies the needs of this group of relatively young adults for social integration. They show exceptionally high viewership of soap operas, religious programs, game shows, and talk shows. If PTV could satisfy the needs of this group for social integration in a manner consistent with its commitment to quality programming, this segment reflects a high potential for shifting its viewing from commercial to public broadcasting.

As indicated by the current viewing behavior, the members of the Home- and Community-Centered segment tend to spend much of their weekday time at home with relatively heavy television viewing. They view less than average amounts of television on weekends.

Their patterns of other media use are distinctive primarily due to heavy newspaper readership, where they are light readers of world news and heavy readers of sections such as gardening, cooking, advertising, social news, and personal advice. Women's service magazines are their favorites.

Elderly Concerns

Members of this segment are especially interested in certain types of informational PTV programming (news and local programs), especially regarding issues that concern the elderly. Their interest would probably be enhanced by a format that offered commentary and analysis, as opposed to mere presentation of the news. These people are also attracted to dramas, soap operas, game shows, religious programming, talk shows, and variety entertainment.

A key factor in gaining viewership from this segment is the ability to provide a continuity and familiarity of format and personalities from show to show. The presence of nonthreatening, familiar hosts appears to satisfy their needs for vicarious participation in a kind of "extended family" and for social integration and acceptance. With this element present, members of the Elderly Concerns segment can be attracted to view a variety of program content areas. The potential for attracting them into the PTV audience should be quite high.

Members of the Elderly Concerns segment are at home and view television well above the average during weekday mornings and afternoons.

They are relatively light users of radio and magazines, but are average in newspaper readership. The sections they read above or near average are news, gardening, and cooking.

Money and Nature's Products

Increased news and local events/issues programs on PTV appear to provide the most promising opportunities for greater viewing by members of this segment. They tend to be oriented toward seeking information. If news and information programming were offered, particularly if it were geared toward financial problems of the elderly (e.g., retirement, social security, health care), it would have an excellent chance of attracting the elderly males in this segment to PTV. These viewers tend to have a politically and socially conservative bent and are also more inclined to favor programming that offers evaluation and analysis rather than pure informational content.

The presentation of such programming would be enhanced by the continued presence of a single authoritative personality, a relatively mature male in his 50s or 60s, who would help to provide the social support and contact that these viewers seek.

Program awareness targeted against this segment can be achieved by advertising in daily newspapers, especially the news, sports, and business and real estate sections. They are relatively light users of radio and magazines (including *TV Guide*).

Notes

1. For a more detailed analysis of our data base, see National Analysts (1981).
2. As indicated in Chapter 2, Blacks were oversampled at the rate of approximately two to one in order to provide a substantial sample size for subsequent analysis.

MINORITY AUDIENCES 207

3. Ideally this analysis would be conducted at the segment level, comparing and contrasting minority and nonminority members within each of the five targeted interest segments. However, given the very small sample sizes of minorities within each of the interest segments, we decided to examine the data for minorities only at the aggregate level.

10

Summing Up

Refining PTV's Objectives

PTV has three types of objectives: functional, quality, and audience reach. With respect to functional objectives, the most commonly used references to define the content of PTV are those contained in the Public Telecommunications Financing Act of 1978, in which mention is made of the use of such media for instructional, educational, and cultural purposes. With respect to the quality of programming, PTV has strived consistently to maintain high standards of professional and artistic excellence. Finally, it has clearly been the intent of Congress, the Carnegie Commission Reports, and the Corporation for Public Broadcasting itself to attempt to provide a diversity of programming designed to reach the entire American population.

The strategy of "audience reach" that is articulated for PTV is fundamentally different from that employed by commercial television. It is presumed that reaching the entire American audience via PTV would be done by building a mix of programs, each of which would have appeal to a particular group, class, or segment within society (many of which might be quite small), but that collectively would ensure that each person in the public was served to some extent by the mix. Commercial television, on the other hand, tries to attract the maximum audience possible for each program broadcast, in an effort to obtain more money for the commercials that accompany it, since the revenue from commercials is pegged to the size of the viewing audience.

As has been discussed in this book and elsewhere, PTV has a long way to go to achieve the goal of serving all of American society. Of the population age 13 and over, 46.5% report never viewing it. Only 35.5% report usually watching once a month or more, while just 25.7% report watching one or more times a week.

PTV has an extremely uneven appeal across the fourteen interest segments. At one extreme, among people in the Arts and Cultural Activities and Cosmopolitan Self-Enrichment segments, about half watch PTV once a week or more. At the other extreme, composed of those in the Elderly Concerns and Family- and Community-Centered segments, fewer than 13% watch this frequently. Only individuals in the Arts and Cultural Activities and Cosmopolitan Self-Enrichment segments came close to watching the full range of programming aired on PTV. These two segments, however, account for only 17% of the U.S. population age 13 and over.

If the gap between the audience PTV serves and that which it desires to serve were not motivation enough to take a closer look at its audience, other factors have made it imperative. Prospects for federal funding appear dim. Due to the advent and expansion of cable television, video cassettes, and videodisks, the range of viewer options is rapidly expanding. In addition, there have been announcements by commercial organizations of pay television efforts aimed at providing arts and cultural programming that will compete with many PTV offerings. These factors, on the one hand, materially increase the difficulty of improving PTV program quality and diversity, while, on the other hand, heighten the need for improved understanding of the motivations behind viewer program choices. In 1974, when we first became involved with the efforts that led to this project, it was argued by numerous professionals that the type of information contained in this volume would be a luxury for PTV, but of little real value. In our opinion, the luxury has now become a necessity.

If PTV is to achieve its audience reach objectives by assembling a diversity of programming that collectively, rather than individually, achieves them without sacrificing the quality of its programming, it is imperative that PTV:

(1) assign more weight to audience-related objectives in determining the mix of programs that are developed and aired; and
(2) explicitly develop those objectives in terms of who is to be served and what interests and/or needs are to be satisfied by the programs developed and aired.

Without some agreed-upon conceptualization of the types of people that comprise the American audience in terms of their interests and needs, it will be difficult to translate the functional and quality objectives of PTV in terms that are meaningful to each of the many and diverse segments of American society. For example, excellence to someone in the Competitive Sports and Science/Engineering segment might be best embodied in programming that emphasizes outstanding athletic accomplishments in competitive sports, whereas the same concept of "excellence" may, if it is to attract members of the Cosmopolitan Self-Enrichment segment, be best embodied in booklike vehicles such as are currently being dramatized on PTV (e.g., *Masterpiece Theatre* and *Dickens of London*).

Implicit in the concept of market segmentation research in general, and in the specific fourteen segments developed and reported on in this book, is a conviction that the audience composition for PTV as a network, and for individual PTV stations, need not, and indeed should not, be the result of pure happenstance. If PTV is to achieve not only its functional and quality objectives but also its audience reach objectives, then we believe it imperative that a conscious and deliberate identification of target audience segments be made for virtually every program selected for broadcast on PTV. Though there are an infinite number of ways to classify audience members into segments, only one has had any continuity of use in the past, namely demographic categories, based primarily on age and sex.

Demographic versus Interest Segmentation

The traditional approach to the study of television audience segments, in both commercial and public television, has been to define them using demographic and socioeconomic characteristics as the dominant frame of reference. Historically, the most frequently used characteristics have been demographic ones, namely, the age and sex characteristics of viewers. More recently, certain ethnic categories, such as Black and Hispanic, have also been used as the basis for defining audience segments. Emphasizing demographic segmentation as a management tool for determining its objectives and programming strategy has a major limitation. While a demographic taxonomy can be helpful in determining *who* should be appealed to, it provides little guidance in determining *what*, by way of general programming strategy or choice of individual programs, is likely to appeal to those segments. Knowledge of the interests and needs of target audience segments, *combined* with their demographic and socioeconomic characteristics, should be maximally helpful to management in determining not only who is to be appealed to, but how to accomplish those objectives.

Much of the historical tradition behind the use of demographic segmentation for media analysis purposes has been based on the fact that the data involved were being developed primarily to help in deciding what media to use for placing advertising to reach target demographic segments. The problem of determining appropriate media content to communicate most effectively with target audiences is sufficiently different in character, we believe, that somewhat different sets of data are appropriate.

In practice, demographic and socioeconomic data are somewhat helpful in answering both the "who" and "why" questions, because such data are, to some extent, correlated with interests and needs. But why settle for what are at best rough indicators of interests and needs, especially if coherent segments, based on direct interest measures, can be identified and shown to be associated with differential PTV viewing and other media behavior?

As we have stated earlier in this chapter, given our belief that the positioning and programming of public television should be the result of a planned process, one can make an extremely strong argument that such a process would be more effective if it were based on an interest segmentation, supplemented with socioeconomic and demographic data such as we have developed.

In summary, our argument is threefold:

(1) The positioning of PTV in general (as well as individual stations and programs), if it is to achieve its functional, quality, and audience reach objectives, should be the result of an explicitly planned process.
(2) The choice of the mix of programs to be given development priorities should begin with the identification of the specific audience segment(s) to which each program in that mix is intended to appeal, along with the intended reach of the entire program mix.
(3) For the aforementioned purposes, an interest-based segmentation scheme (supplemented by demographic and socioeconomic measures) is considerably more appropriate than one defined solely by traditional demographic and socioeconomic descriptors. Surely, such descriptions as those characterizing the Cosmopolitan Self-Enrichment, the Family- and Community-Centered, and the News and Information Segment members comprise far richer inputs to understanding what will appeal to them than do their counterparts based primarily on whatever demographic factors one might choose to use.

The principal purpose of this study has been to develop a segmentation scheme that is provocative and stimulating—one that will be useful to those with programming as well as fund-raising responsibilities in implementing *whatever* policy or strategy decisions are eventually adopted by PTV. In the concluding section of this chapter we report our observations as to the prospects for further audience development for each segment. Before moving to this discussion there is, however, one other issue related to audience segmentation that we feel is especially important for those in PTV to confront, namely, the potential conflict between segmentation in the world of politics and that in the world of PTV.

Segmentation in the Worlds of Politics and PTV

People who live near each other tend to be considerably more similar to each other in race, income, education, religion, and other demographic characteristics than those who live far apart. By the very nature of the political process in our country, political constituencies consist of groups of voters who live in geographically contiguous communities. Hence, it is not surprising that many public action groups and many politicians see themselves as needing to be especially responsive to the interests of one or more demographically defined groups in the population—especially when it comes to influencing legislation at the local or state, as well as the federal, level. As appropriate as this process may be for influencing legislative decisions, it has the potential for creating special problems for PTV that may inhibit the development of programming strategies effective in achieving its objectives of excellence and service to the entire American public.

The nub of the possible problem, as we see it, is the tendency for organizations largely dependent on public sector funding to adopt the same or similar frames of reference as do politicians when it comes to addressing their audiences—namely, a demographically driven frame of reference. As we have demonstrated through the findings in this book, as well as in our previous book, *The Public's Use of Television,* these demographic categories contain people who are extremely heterogeneous with regard to their interests, their needs, and their media habits. We believe it is dysfunctional for PTV to respond to moral suasion to employ in its planning a demographic segmentation framework that may well be appropriate for political purposes, but much less effective in conceptualizing and achieving its objectives.

It is imperative for PTV to have, and to exercise, its freedom to study the interests and needs of the American audience and to segment that audience in terms that are sensible and useful for the purposes of developing programming. As we have indicated, this process is bound to lead to a segmentation scheme of the American audience that is fundamentally different from that used in the world of politics or in the world of media selection. However, if PTV is successful in achieving its goal of serving all or most of the entire population (i.e., all fourteen interest segments), it will, in doing so, serve all demographic and socioeconomic segments as well. Furthermore, it will achieve this end through a process that is less politicized and less adversarial in nature. For PTV's purposes, our interest scheme may well not be perfect; but it is definitely more appropriate for making many programming decisions than the more traditional, demographic frame of reference.

PTV Audience Development

If PTV is to make material gains in broadening its audience coverage, while at the same time maintaining the standards of excellence that has characterized its past programming, it must follow a strategy of special interest narrowcasting in determining the mix of programs it offers.

Only two segments, accounting for 17% of the U.S. population age 13 and over, demonstrate an unambiguous attraction to abstract, culturally upscale material, almost irrespective of the special interest content area it represents. These are the people in the Arts and Cultural Activities and Cosmopolitan Self-Enrichment segments. With the exception of these two segments, one observes extremely modest levels of PTV viewing and finds program viewing patterns that reflect more limited interests in subject matter not driven by the generalized attraction to abstract, culturally upscale material. Excellence embedded in this type of programming has a limited potential, in and of itself, to attract voluntary viewing.

For artistic excellence to attract viewers from the remaining twelve segments, it needs to be coupled to functional benefits that are meaningful to the individuals in the particular segment or segments selected as audience targets. For example, if one wishes to attract viewers in the Family-Integrated Activities segment, it is useful to understand their desires to guide their children's intellectual development. With this knowledge it is possible to develop programs that lend themselves to shared viewing between such parents and their children, and to create through advertising and promotion an awareness of this benefit among members of the target segment. As *Sesame Street* amply demonstrates, though the task is difficult, it is nonetheless achievable, providing one understands the benefits sought by viewers and finds ways to deliver them. For yet other people, such as those in the News and Information segment, a more detailed understanding of the role news plays in their everyday socialization is no doubt a useful prerequisite to determining what, if any, additional programming vehicles can serve their needs and yet meet the same standards of excellence as does the *McNeil/Lehrer Report*.

As long as PTV maintains the standards it has in the past in its efforts to increase viewing among people in the other twelve segments, it is likely that they will be able to maintain or increase the current levels of usage among individuals in its two heavy viewing segments. One should not take for granted an existing audience as one strives to attract yet more viewers. However, the generalized attraction of people in both of these segments to culturally upscale materials is likely to cause them to maintain, if not increase, their usage of PTV irrespective of the special interest areas that PTV may pursue in its efforts to attract viewers from other segments.

Of the remaining twelve segments there is one for which we think the potential for increased PTV viewing is virtually nil, namely those in the Detached segment, accounting for 9% of the population aged 13 and over. Their overall usage of all media is relatively low and they show little evidence of any felt needs for intellectual stimulation or creativity. If there is a way to attract them to PTV, they have not provided us with relevant directions in the information we collected from them.

The eleven segments that comprise the remaining 74% of the U.S. population suggest an extremely diverse set of opportunities for PTV program development. In the discussion that follows, we briefly illustrate some strategies that we think might attract additional viewers from each of them. A more detailed discussion of their characteristics and audience potential is interwoven throughout Chapters 4 to 9. What follows is meant to serve as a catalyst to the thinking of others and not as a comprehensive list of possibilities. The eleven segments are discussed in the order in which they were initially introduced in Chapter 3.

Adult Male Concentration

Mechanics and Outdoor Life. Much of the media use of individuals in this segment is associated with their needs to escape or to fantasize. PTV can respond to them by offering quality programming consistent with their interests and needs. For example, there are many excellent literary works in the areas of science fiction and crime drama. Properly scripted and, perhaps more importantly, properly advertised (e.g., emphasizing their *excitement* rather than their literary merit or historical interest), such programming could be successful in reaching these people.

Other elements of program content that might have appeal to the people in this segment are those associated with away-from-home activities that emphasize personal development or physical activity such as camping out, hiking, jogging, or biking.

Money and Nature's Products. The people in this segment appear to respect experience and success as well as value exposure to people who have attained them. This is especially true if the successful personality is a mature male, whose achievements reflect what this segment perceives to be solid, traditional American values.

Programming related to business or national and local affairs might well attract additional members of this segment into the PTV audience. This would be particularly likely if the program content focused on prominent people and how they were approaching and/or coping with significant problems of interest to this group of viewers.

This same general tactic of providing in-depth coverage of an issue or an unfolding event from the perspective of a significant personality with direct, immediate involvement might also be extended to the outdoor interests of these people. Outdoor-related material is more apt to be interesting to this segment if seen through the eyes of an experienced professional such as a fisherman or hunter.

Home- and Community-Centered. This segment's need for social integration during the day is similar to that of the Elderly

Concerns segment. During this period, the Home- and Community-Centered segment can be attracted by programming similar to that recommended for the Elderly Concerns segment—religious programs and programs that facilitate a sense of personal identification and participation in an adult world.

They can also be reached through programming that focuses on local community personalities, activities, and events. In this area, local PTV stations have the opportunity to augment the coverage of local newspapers, providing greater depth and a sense of vicarious participation. It is likely, for example, that a talk show with local guests, a cooking program featuring recipes with a regional slant, or a gardening program talking about local flora would attract these people.

Adult Female Concentration

Elderly Concerns. Virtually none of the 22 PTV programs included in the study address this segment's interest in religious content, or their needs to cope with loneliness. *Women* and *Upstairs, Downstairs* are as close as PTV has come to dealing with these latter needs. Had *Over Easy* been included it would have probably fared even better in reaching these people than the above programs.

It is also clear that to reach these people more often PTV must feature individuals and social contexts (e.g., family settings) that are not only nonthreatening, but designed to facilitate a sense of personal identification and participation. Continuity of characters and themes can be extremely important elements in achieving these goals. Religious content offers another means of attracting them. As is clear from the wide variety of commercial television programs they view, there exists a wide range of program formats that might increase their involvement with PTV.

Family- and Community-Centered. It is quite likely that PTV has at least one major appeal to the people in this segment; namely, the presence of little or no objectionable material. By the same token, given their family and community interests, there is little on PTV that is tailored to serving their needs.

There are excellent children's programs on PTV. However, individuals in this segment have children who are, on average, three years older than those in the Family-Integrated Activities segment, which reports high levels of PTV viewing.

Certainly, more attention to programming with religious themes might attract more viewers from this segment, as might programs emphasizing the developing and sharing of traditional values within families—especially families with adolescents or early teens. Finally, programming focused on local community activities might, and possibly already does, attract people in this segment. Our study did not measure the viewing of such local programs and, therefore, may underestimate the overall PTV usage of this segment.

Family-Integrated Activities. Television viewing as well as other media behavior on the part of people in this segment is more influenced by the presence of children than that for any other segment. It is clear that the viewing behavior of the adults in this segment is largely associated with programs aimed either toward the intellectual and social development of young children, or is focused on family interaction patterns. Through their selection of programs they guide their children's development and share the viewing experience with them. The continued development of children's programming on PTV is one obvious strategy for maintaining or increasing the amount of PTV viewing that emanates from this segment.

A second strategy for serving the needs of viewers in this segment is one aimed more directly at serving their own personal needs. This strategy would emphasize programs that provide vehicles for improving one's effectiveness, not only in guiding the intellectual and social development of one's children, but also in fostering a closer, more supportive set of nuclear family relationships rewarding to all of its members. Such a programming strategy could take a wide variety of formats. For example, educational programs (including talk shows) could be developed that would focus on child development and family life. In addition, one might also develop dramas that concern themselves with the same issues, but that focus on ways nuclear families provide direction and support to their members as they face life's problems.

Youth Concentration

Competitive Sports and Science/Engineering and *Athletic and Social Activities*. A strong argument can be made that the least-recognized and catered to minority audience in PTV is not children, Blacks, or even Hispanics, but adolescents. These people, who are no longer children but have not yet developed the interest and need patterns that characterize the adult segments, seem to "fall between the cracks" when it comes to PTV programming strategy. This is true not only of people in these two segments, but also characterizes a substantial proportion of those in the Indoor Games and Social Activities segment. While members of the latter segment are on average a bit closer to adulthood, they nevertheless exhibit many needs similar to those of the younger two segments. Individuals in all three of these segments generally avoid the kind of intellectual, cultural, informational or abstract material that constitutes much of what is currently aired on PTV.

One means of increasing PTV usage by people in the first two segments and those in the Indoor Games and Social Activities segment, many of whose members are still in school, is the development and integration of programming related to high school curricula. The creation of visually dramatic educational material related to standard science, history, or literature courses could stimulate viewing by these segments, particularly if strongly encouraged or required by the schools. PTV might even consider supporting the development of classroom materials, such as texts, workbooks, examinations, or other presentation materials to complement the use of PTV programs in a classroom or home setting. Another approach might attempt to develop programming that combines education and entertainment in a manner consistent with the everyday life of people (especially those in school) in these segments. This could be the equivalent of a *Sesame Street* aimed at adolescents, rather than young children.

Heavy use of parody or satire, for example, might be an effective means of combining entertainment with educational material.

Substantial increases in PTV exposure among these Youth Concentration segments are not likely to be achieved unless PTV seeks innovative ways of attracting them. Understanding their patterns of interests and needs, however, provides a basis for trying to reach them, should PTV choose to do so.

In addition to these general strategies there is clearly potential for attracting more members of the Competitive Sports and Science/Engineering segments to PTV by increased sports programming. Their appetite for the subject appears to be almost insatiable. Such programming need not be restricted to the broadcasting of live events, but could incorporate presentations of the history of sports, biographical material on famous athletes, discussions of strategies for winning, and other related topics.

Indoor Games and Social Activities. While this segment contains people young enough to share the adolescent-oriented needs discussed in the previous section, it also contains many who have entered the world of adulthood, some with parental responsibilities. It should be recalled that this segment is predominantly female and that it contains a much higher percentage of individuals with young children than do the other two segments.

The Indoor Games and Social Activities segment would probably be attracted into the PTV audience by programming designed to help young women who are learning the role of homemaker to adjust to that role. Programs could provide information and instruction in an entertaining manner on cooking, gardening, housekeeping, and related subjects. Note, however, that in order to appeal to this segment, such programs would have to acknowledge their youth and inexperience. While these young women may want to learn to cook, they are probably not ready for gourmet training.

They are also available during the day for programming that provides an alternative to soap operas and game or quiz shows. During this period, they might respond to programs that provide alternative forms of entertainment incorporating either contemporary music or comedy, since these elements appear to be present in much of their current television viewing. The development of such programs on PTV, in order to be successful in attracting this segment, should feature young adults, especially women, and should deal with topics and issues of special concern to people of this age group, without appearing to be pedantic.

This segment ranks fourth in the amount of television they watch, but only ninth in their viewing of PTV. Consequently, their potential for increased viewing of PTV is significant should they be identified as a target audience segment and efforts be made to reach them. Their relative youth adds to their attractiveness, as once they become familiar with and form the habit of viewing PTV, they are likely to remain in the audience for years to come, even as their interests, needs, and life styles change and they shift into other audience segments.

Mixed

News and Information. The single most all-pervasive characteristic of people in this segment is their desire to keep informed on a broad range of subjects. This need is reflected, not only in their commercial television and PTV usage, but also in their readership of magazines and newspapers. The knowledge they seek appears to serve their needs to be more socially stimulating and better able to converse with others. Continued or increased emphasis on news programming, talk shows, and documentaries would appear to be the principal means of maintaining or even increasing audience share in this segment.

Highly Diversified. This segment derives its name from the fact that its members are interested in an unusually broad range of subjects, and correspondingly use media, especially television, as a means of pursuing those interests. They possess an unusually high need for intellectual stimulation, especially if one takes into account their educational level, which is well below that of other segments demonstrating a similar need.

Because of the unusually broad range of their interests and media usage, it is likely that the reach of PTV among members of this segment will be maintained or even increased with programming strategies designed to attract members of the other segments. Their attraction to PTV appears to be based more on the quality of execution and the intellectual stimulation provided by PTV programming in general than on the specific nature of the program content itself.

An additional device for developing this segment as an audience would be the use of Black or Hispanic actors, formats, and themes. This segment's membership is disproportionately composed of people in these two minority groups, and both tend to be attracted to programs whose characters or themes are easier for them to relate to.

In Conclusion

As the recommendations in the preceding section imply, we are convinced that it is feasible for PTV to extend its reach to more members of society without sacrificing the standards of excellence that it has pursued in the past. The large majority of the U.S. population is not attracted to abstract, culturally upscale material in any medium, let alone television, which is treated by most as a vehicle primarily for entertainment. Nonetheless, given the needs and interests of every segment except one, there are one or more aspects of their interests and needs that are potentially addressable by PTV without any inherent conflict with its pursuit of excellence. The attraction of this diverse mix of audience segments implies the need for "narrowcasting," but with a greater variety of program types than has previously been employed by PTV. In addition, it implies a program development and selection process that, from the outset, includes explicit objectives defined in terms of who is to be reached and what interests and needs are to be served in reaching them. The information contained in this volume is designed to be helpful to those concerned with PTV's efforts to develop a broader audience, whether they be authors, programmers, legislators, regulators, or concerned citizens.

By no means do we view our particular set of fourteen interest segments as the only appropriate model for analyzing potential television audiences. Patterns of interests change, needs change, and media behavior changes in response to numerous forces in our society. Consequently, any given segmentation scheme carries the seeds of obsolescence in the data base from which it is constructed.

Even with the present data base, we recognize that some potential users may not be totally comfortable with the particular structure (the fourteen interest segments) presented here. Others may wish to employ a structure with fewer or more interest segments or with different ones. They may want to relate other data from the study to the basic segment structure. With this in mind, we are making the entire data base available to anyone who wishes to analyze it using other approaches.[1]

In conducting this research and reporting its results, our goal has been to demonstrate the potential value of employing an interest-based audience segmentation scheme to provide a conceptual framework for strategic planning in the area of PTV. We will consider ourselves successful if we have provoked others to begin to think about PTV audiences in other than traditional demographic terms and if we have stimulated researchers to undertake further studies that attempt to gain insight into the psychological correlates of television viewing in general and PTV viewing in particular.

Note

1. A complete data tape and detailed documentation, along with copies of the questionnaire, can be obtained at a current (1982) cost of $350.00 by writing to:

> National Analysts
> 400 Market Street
> Philadelphia, PA 19106

Availability and cost are subject to change in future years.

Table A.1 Percentage of Persons Watching by Program and Interest Segment

	Entire Population	Arts and Cultural Activities (AF)[a]	Cosmopolitan Self-Enrichment (M)	News and Information (M)	Highly Diversified (M)	Family-Integrated Activities (AF)	Athletic and Social Activities (Y)	Home- and Community-Centered (AF)	Mechanics and Outdoor Life (AM)	Indoor Games and Social Activities (Y)	Competitive Sports and Science/Engineering (Y)	Detached (M)	Money and Nature's Products (AM)	Elderly Concerns (AF)	Family- and Community-Centered (AM)
Documentaries															
Age of Uncertainty	1.5	5.0	3.3	2.4	1.8	1.0	.2	—	1.2	.7	3.1	.3	.4	1.3	—
Nova	4.4	12.0	12.2	4.4	3.5	6.5	.5	2.4	1.8	.8	2.8	1.7	2.3	3.3	1.1
Theatrical Performances															
Dickens of London	3.2	15.3	13.0	1.2	.8	—	1.1	2.4	1.2	2.0	.4	1.1	2.8	.5	.1
Visions	1.5	8.1	3.7	2.3	.9	—	.7	1.2	—	1.6	.3	1.3	—	.2	.1
Upstairs, Downstairs	3.2	12.7	4.9	1.0	2.5	.3	.2	.5	2.6	1.3	.5	1.9	3.1	5.4	4.6
In Pursuit of Liberty	.8	2.4	.3	3.5	.8	.7	.2	1.0	—	.7	—	.3	.4	.4	.5
Great Performances	4.1	18.2	11.9	4.7	5.1	.4	.3	.6	.4	1.5	—	1.1	4.0	3.7	1.0
The Best of Families	3.0	7.3	3.5	6.1	3.8	5.6	.2	.3	1.4	3.9	1.0	2.4	1.4	2.6	—
Masterpiece Theatre	6.2	20.7	11.0	6.1	3.1	6.9	.8	4.0	2.3	3.1	3.8	1.7	5.8	5.0	6.8
Once Upon a Classic	3.4	6.8	14.4	1.6	3.7	5.2	2.2	.3	.8	1.9	1.2	1.7	2.0	.7	1.0
Musical Performances															
Evening at Symphony	7.0	32.0	16.7	2.3	2.7	2.7	—	4.7	4.4	2.1	.9	2.5	8.5	4.8	4.0
Opera	4.5	28.0	6.3	.6	2.7	3.7	.2	2.6	.2	1.2	—	2.1	4.0	2.7	.4
Evening at Pops	6.1	22.5	21.9	4.4	2.9	2.1	—	3.5	2.6	1.7	1.1	2.9	4.5	4.0	4.0

(continued)

Table A.1 Continued

	Entire Population	Arts and Cultural Activities (AF)[a]	Cosmopolitan Self-Enrichment (M)	News and Information (M)	Highly Diversified (M)	Family-Integrated Activities (AF)	Athletic and Social Activities (Y)	Home- and Community-Centered (AF)	Mechanics and Outdoor Life (AM)	Indoor Games and Social Activities (Y)	Competitive Sports and Science/Engineering (Y)	Detached (M)	Money and Nature's Products (AM)	Elderly Concerns (AF)	Family- and Community-Centered (AM)
News/Documentaries															
Washington Week in Review	4.9	13.8	4.7	10.3	5.6	.9	—	5.2	2.2	.4	1.0	2.0	10.2	4.5	5.7
Wall Street Week	2.8	11.5	2.5	5.8	1.8	—	—	2.0	3.0	.7	—	1.3	4.8	—	4.5
Black Perspective on the News	2.5	2.7	1.6	6.1	7.3	1.0	.5	2.8	2.2	1.6	.5	2.4	.6	3.5	1.5
MacNeil/Lehrer Report	3.1	8.7	8.8	11.8	4.1	—	—	2.4	—	.9	—	1.5	2.4	.8	1.6
Children's Programs															
Sesame Street	5.8	3.2	11.2	4.6	11.7	17.0	5.4	6.8	3.0	3.5	2.6	2.4	1.3	.9	1.4
Mister Rogers	2.0	.2	3.6	—	2.3	8.4	4.7	2.8	—	1.2	1.4	.4	.4	—	.4
Electric Company	4.1	2.2	9.5	3.7	5.1	11.2	4.7	4.6	3.8	2.4	2.0	1.9	.6	.7	.6
Other															
Women—talk show	.9	.7	.6	2.1	1.6	.8	—	.3	.2	1.3	—	2.1	.3	2.1	.7
The French Chef	1.5	3.1	2.9	2.2	2.7	1.1	.6	2.2	—	1.2	.4	1.7	.6	.9	.7
Grand Average	4.6	10.9	7.7	4.0	3.5	3.4	1.0	2.4	1.7	1.6	1.0	1.7	2.7	2.2	1.9

a. Letters associated with each segment indicate which concentration it is in, namely: AF = Adult Female, AM = Adult Male, Y = Youth, and M = Mixed.

Table A.2 Viewing Frequency Ratios by Program and Interest Segment

	Entire Population	Arts and Cultural Activities (AF)[a]	Cosmopolitan Self-Enrichment (M)	News and Information (M)	Highly Diversified (M)	Family-Integrated Activities (AF)	Athletic and Social Activities (Y)	Home- and Community-Centered (AF)	Mechanics and Outdoor Life (AM)	Indoor Games and Social Activities (Y)	Competitive Sports and Science/Engineering (Y)	Detached (M)	Money and Nature's Products (AM)	Elderly Concerns (AF)	Family- and Community-Centered (AM)
Documentaries															
Age of Uncertainty	.04	250	150	275	150	75	–	–	25	25	250	25	–	125	–
Nova	.10	240	200	50	100	150	10	90	30	10	80	60	50	90	30
Theatrical Performances															
Dickens of London	.09	522	322	44	144	–	33	78	22	33	11	33	44	11	–
Visions	.03	400	200	268	133	–	33	133	–	100	–	100	–	–	–
Upstairs, Downstairs	.10	470	190	30	80	10	–	20	30	40	–	50	80	130	140
In Pursuit of Liberty	.02	200	50	700	200	150	–	150	–	50	–	50	–	50	50
Great Performances	.09	478	200	144	144	11	11	22	11	56	–	22	111	89	33
The Best of Families	.07	229	114	314	171	114	–	–	43	129	14	100	51	100	–
Masterpiece Theatre	.16	369	150	69	56	113	6	69	44	75	50	25	81	94	88
Once Upon a Classic	.08	188	375	50	125	188	75	–	38	25	25	50	50	–	38
Musical Performances															
Evening at Symphony	.13	462	200	38	38	38	–	77	69	38	8	46	115	92	15
Opera	.09	589	122	11	111	78	–	89	–	44	–	44	67	67	–
Evening at Pops	.12	375	242	117	75	42	–	58	50	42	17	75	58	58	42

(continued)

Table A.2 Continued

	Entire Population	Arts and Cultural Activities (AF)[a]	Cosmopolitan Self-Enrichment (M)	News and Information (M)	Highly Diversified (M)	Family-Integrated Activities (AF)	Athletic and Social Activities (Y)	Home- and Community-Centered (AF)	Mechanics and Outdoor Life (AM)	Indoor Games and Social Activities (Y)	Competitive Sports and Science/Engineering (Y)	Detached (M)	Money and Nature's Products (AM)	Elderly Concerns (AF)	Family- and Community-Centered (AM)
News/Documentaries															
Washington Week in Review	.13	262	77	269	108	15	–	131	38	–	8	54	269	69	100
Wall Street Week	.07	357	57	343	57	–	–	100	86	14	–	43	143	–	114
Black Perspective on the News	.07	100	71	257	343	29	29	114	43	71	29	114	14	129	51
MacNeil/Lehrer Report	.11	277	208	400	138	–	–	77	–	8	–	31	46	15	46
Children's Programs															
Sesame Street	.21	57	210	76	148	333	110	119	19	62	48	43	19	14	24
Mister Rogers	.07	14	214	–	129	514	186	114	–	29	100	29	–	–	14
Electric Company	.13	54	185	77	146	338	123	115	46	54	69	38	8	8	23
Other															
Women—talk show	.03	67	33	133	133	67	–	33	–	33	–	200	33	333	67
The French Chef	.04	150	225	75	250	75	50	125	–	50	50	125	50	25	50

b. Letters associated with each segment indicate which concentration it is in, namely: AF = Adult Female, AM = Adult Male, Y = Youth, and M = Mixed.

References

BERSTEIN, P. W. (1979) "Television's expanding world." Fortune (July): 64-69.
BLUMER, J. G. and E. KATZ (1974) The Uses of Mass Communication. Beverly Hills, CA: Sage.
BOWER, R. T. (1973) Television and the Public. New York: Holt, Rinehart & Winston.
Carnegie Commission on Educational Television (1967) Public Television: A Program for Action. New York: Harper & Row.
Carnegie Commission on the Future of Public Broadcasting. (1979) A Public Trust. New York: Bantam Books.
CATER, D. and M. J. NYHAN [eds.] (1976) The Future of Public Broadcasting. New York: Praeger.
COMSTOCK, G. (1975) Television and Human Behavior: The Key Studies. Rand Report R-1747-CF. Santa Monica, CA: The Rand Corporation.
——— and M. FISHER (1975) Television and Human Behavior: A Guide to the Pertinent Scientific Literature. Rand Report R-1746-CF. Santa Monica, CA: The Rand Corporation.
COMSTOCK, G. and G. LINDSEY (1975) Television and Human Behavior: The Research Horizon Future and Present. Rand Report R-1748-CF. Santa Monica, CA: The Rand Corporation.
COMSTOCK, G., S. CHAFFEE, N. KATZMAN, M. McCOMBS, and D. ROBERTS (1978) Television and Human Behavior. New York: Columbia University Press.
Corporation for Public Broadcasting (1980) Proceedings of the 1980 Technical Conference: Quantitative Television Ratings. Washington, DC: CPB Office of Communication Research.
——— (1978) A Quantitative Study: The Effect of Television on People's Lives, Vol. 1. Washington, DC: CPB Office of Communication Research.
——— (1976) Mission and Goals—Tasks and Responsibilities. Washington, DC: CPB Office of Communication Research.
FRANK, R. E. and M. G. GREENBERG (1980) The Public's Use of Television: Who Watches and Why. Beverly Hills, CA: Sage.
——— (1976) "Audience segmentation analysis for public television program development, evaluation and promotion." The John and Mary R. Markle Foundation, Philadelphia. (unpublished)
FRANK, R. E. and W. F. MASSY (1975) "Noise reduction in segmentation research," in John U. Farley and John A. Howard (eds.) Control of "Error" in Market Research Data. Lexington, MA: D. C. Heath.
——— and Y. WIND (1972) Market Segmentation. Englewood Cliffs, NJ: Prentice-Hall.

FUNT, P. (1979) "Broadcasters are switching to narrowcasting." New York Times (December 16): D43.
HOWARD, N. and B. HARRIS (1966) A Hierarchical Grouping Routing FORTRAN IV Program. Philadelphia: University of Pennsylvania Computer Center.
KAPLAN, M. (1975) Leisure: Theory and Policy. New York: John Wiley.
KATZ, E. (1977) Social Research on Broadcasting: Proposals for Further Development. London: British Broadcasting Corporation.
——— M. GUREVITCH, and H. HASS "On the use of the mass media for important things." American Sociological Review 38 (April): 164-181.
LYLE, J. (1975) The People Look at Television: 1974. Washington, DC: CPB.
MORRISET, L. N. (1976) "Rx for public television," pp. 163-184 in D. Cater and M. J. Nyhan (eds.) The Future of Public Broadcasting. New York: Praeger.
MENDELSOHN, H. (1966) Mass Entertainment. New Haven, CT: College and University Press.
MYRICK, H. and C. KEEGAN (1981) Review of 1980 CPB Communication Research Findings. Washington, DC: Corporation for Public Broadcasting Office of Communication Research.
National Analysts (1981) Attracting Minority Audiences to Public Television. Washington, DC: Corporation for Public Broadcasting Office of Communication Research.
——— (1975) "Toward the identification of special interest audiences for public television." The John and Mary R. Markle Foundation, Philadelphia, March. (unpublished)
NIGEL, H. and B. HARRIS (1966) A Hierarchical Grouping Routing FORTRAN IV Program. Philadelphia: University of Pennsylvania Computer Center.
ROBINSON, G. O. (1978) Communications for Tomorrow. New York: Praeger.
STEINER, G. A. (1963) The People Look at Television. New York: Alfred A. Knopf.
WELLS, W. D. [ed.] (1974) Life Style and Psychographics. Chicago: American Marketing Association.

HE 8700 .7 .A8 F69

AUG 4 '87